Algorithms for Games

G.M. Adelson-Velsky
V.L. Arlazarov
M.V. Donskoy

Algorithms for Games

With 38 Illustrations

Springer-Verlag
New York Berlin Heidelberg
London Paris Tokyo

G.M. Adelson-Velsky, V.L. Arlazarov,
and M.V. Donskoy
Institute for Systems Studies
9, Prospekt 60-Let Oktyabria
117312 Moscow B-312, USSR

Translator
Arthur Brown
P.O. Box 326
Garrett Park, MD 20766, USA

Library of Congress Cataloging-in-Publication Data
Adelson-Velsky, G.M. (Georgiĭ Maksimovich)
Algorithms for games.
Bibliography: p.
Includes index.
1. Game theory. 2. Algorithms. I. Arlazarov,
V.L. (Vladimir L'vovich) II. Donskoy, M.V.
III. Title.
QA269.A35 1988 519.3 87-23549

Translation of the original Russian edition
Programmirovanie Igr. © 1978, Nauka, USSR.

Text prepared by the translator using the Spellbinder® word processing software
and the STI Math Program.
Typeset by Science Typographers, Inc., Medford, New York.
Printed and bound by R.R. Donnelley and Sons, Harrisonburg, Virginia.
Printed in the United States of America.

9 8 7 6 5 4 3 2 1

ISBN 0-387-96629-3 Springer-Verlag New York Berlin Heidelberg
ISBN 3-540-96629-3 Springer-Verlag Berlin Heidelberg New York

Contents

Preface

Technicians, economists, industrial managers, and many other specialists often need to find an element belonging to a finite set and having certain given properties (provided, of course, that such an element exists). In principle, this problem can be solved by searching the set, element by element. There is, however, a well-rooted belief that this possibility is purely theoretical because the work involved in the search is enormous. Moreover, any general method (the search method included) may turn out to be either good or bad, depending on the concrete circumstances and the method chosen for implementation.

Suppose, for instance, that we need to find a root of the equation $x^4 + x^3 - 1 = 0$ on the interval $(0,1)$ to within an accuracy of 0.1. Then it is better to evaluate the polynomial on the left side of the equation for the values $x = 0, 0.1, \ldots, 1.0$ than to use the exact Ferrari method, in which we first reduce the equation to the form $y^4 + py^2 + qy + r = 0$ by the substitution $x = y - 1/4$ and then transform this equation to the soluble form

$$\left|\left(y^2 + \frac{p}{2} + \alpha\right)^2 - 2\alpha\left(y - \frac{a}{4\alpha}\right)^2\right| = 0$$

where α is a root of the cubic equation

$$q^2 - 8\alpha\left(\alpha p + \alpha^2 + \frac{p^2}{4} - r\right) = 0,$$

after which we bring it into the form $\beta^3 + p'\beta + q' = 0$ by the substitution $\alpha = \beta - p/3$. Finally we use Cardan's formula

$$\beta = \sqrt[3]{\frac{-q'}{2} + \sqrt{\frac{q'^2}{4} + \frac{p'^3}{27}}} + \sqrt[3]{\frac{-q'}{2} + \sqrt{\frac{q'^2}{4} - \frac{p'^3}{27}}}\,.$$

In essence, a search amounts to the solution of problems arising from a given one when the value of an unknown parameter is fixed in one way or another, and a choice is made among a set of contemplated values that yields the most suitable solution. Often each of the contemplated problems, with the parameter fixed, is solved by a search. Then we speak of a multi-level or hierarchic search. If we adopt the most general definition of a search we can give only trivial recommendations for its implementation (though even these may be useful). Fortunately, the kinds of problems we are dealing with often have features in common aside from their solubility by searches. This permits us to establish, study, and apply general search methods.

This book aims to provide a concrete example of the programming of a two-person game with complete information, and to demonstrate some of the methods of solution; to show the reader that it is profitable not to fear a search, but rather to undertake it in a rational fashion, make a proper estimate of the dimensions of the 'catastrophe', and use all suitable means to keep it down to a reasonable size. The game programming problem would seem to be ideally suited to the study of the search problem, and in general for multi-step solution processes. The clarity and relative simplicity of the rules and the scoring of the results, the availability of suitable experimental methods for comparing various solution algorithms (including experiments on human approaches that arrive at answers produced by informal methods), all act to yield suitable methods for developing and trying out different approaches to the solution of problems by search methods.

A hierarchic search underlies all natural methods for finding a move in a game position. 'If I do this, I reach a position where various possibilities are open to me, but my opponent can answer thus and thus...'—this is a typical basis for the choice of a move. Clearly, it involves a search. As we shall show in the first chapter, however, a so-called *full-width* or *exhaustive* search need not require a complete enumeration of all possible positions that might arise by application of the rules of the game. We have in fact dedicated this book to the study of methods for limiting the extent of a search.

We have written about the programming of games (i.e. we have assumed that the reader would like to write a program that would play a game, serious or not) because we believe this is the best way to bring out the ideas and methods we are attempting to expound. We do not, however, pay attention to the technical problems of programming, essential though these may be. There is no specific technique for programming computers to play games. The fact that many technical paradigms were first conceived in the development of game programs does not contradict this assertion. These paradigms have later found wide development and application in the construction of many programs that are used in many different areas. Their prevalence attests to the fact that many clever programmers have been at work on them, and to the fact that they have stretched the capabilities of their machines to the limit.

We shall study games between two opponents named White and Black. Without loss of generality we may assume that these are zero-sum games. They permit at each stage a finite number of moves, for each of which the permissible replies of the opponent are known. Once the opponent's decision is made, a new position arises and is uniquely defined. Some positions are terminal: no decision is allowed to one of the opponents, the other wins and the one to move loses. If a non-terminal position arises, however, the opponent who has the move must in fact move.

We shall study a narrower class of games–namely games with complete information. In every such game we know which opponent is to move at each turn, and he must move. Thus both opponents know what the resulting position is. The positions in a game with complete information fall into three types: White to move, Black to move, and terminal positions. [We might define the notion of a game with several players and complete information, and carry over or suitably transform some of the results obtained in the book, but we shall not take the time to do so.]

Every game begins with a position which from now on we call the base position. In some games, for example card games, the base positions will vary, perhaps depending on some chance event. In other games, as for example chess, the base position is fixed once and for all. Even in chess, the player's concern is not with this fixed base position but rather with positions that arise either in the course of play or, as in problems and endgame studies, by artifice.

If the base position is prescribed, we may construct a game tree. Its nodes correspond to positions, and every arc leads from one position to another that can arise from it by a legal move. The base position corresponds to the root of the tree. A two-person game with complete information can be formally defined by prescribing the game tree, i.e. specifying the color of non-terminal nodes (designating the person who has the move in the given position) and the score corresponding to the just concluded move. These definitions are given in the first chapter and are widely applied throughout the book. To study the concrete properties of a game, however, we shall deal with positions and moves rather than with the game tree.

In Chapter 1 we develop the branch-and-bound method (the α,β-heuristic). We pay primary attention to the theoretical foundation of the method and to an estimate of the minimum size of the search required for its solution. In some places we relinquish an elegant inductive proof in favor of a more unwieldy one, in an effort to isolate the conditions needed for the existence of an objective score for the base position and for finding the best strategy for each of the opponents.

Chapter 2 is devoted to heuristic (i.e. inexact) methods for choosing a move in a contemplated position. We pay special attention to the problems of establishing such methods, and to a discussion of those properties of specific games that ensure good results when the methods are applied. It is worth noting that we employ a probabilistic approach in establishing our heuristic methods for programming games.

In Chapter 3 we develop the theory of analogical reasoning as a basis for decision-making. This theory is founded on the concept of a move that is independent of the position in which it is made; that is, the same move can be made in many different positions. The sequels to such a move may vary, but in many cases the move leads to roughly identical changes. Suppose that in the position B move Ψ is being studied. We want to know what conditions are sufficient to yield the same score for this move in another position C. We formulate these conditions and prove their sufficiency. They amount to this: the sequence of moves leading from B to C must have no influence on the variation that establishes the score for the move under study. In other words, the sequence must consist of moves that have no relationship to the decisive variation.

In Chapter 4 we take up the probabilistic approach to game programming. This approach has four aspects: a) the methods for formulating the elementary stochastic hypotheses and calculating the probability of correctly scoring a given position and finding the best moves; b) the methods for statistical testing of our hypotheses; c) the construction of more effective methods for computing the score and finding the best moves in a given position, on the basis of an analysis of a stochastic model of the game; and d) the probabilistic approach to the programming of games with complete information.

Since the probabilistic approach to game programming has only recently been applied, many of the results obtained in Chapter 4 must be regarded as preliminary, and some of them have not even been established—e.g. the statistical testing of the stochastic hypotheses and the basis of the probabilistic approach. Nevertheless, the results that have been obtained do in our opinion support the prospects for this new direction in game programming and we have therefore included the chapter.

At the end of the book we present an appendix containing a brief sketch of the work that has been done on algorithms for games and a bibliography which includes some references not cited in the text.

Chapters 3 and 4, which contain new results on two-person zero-sum games with complete information, are mutually independent. Formally they are also independent of Chapter 2, but we nevertheless recommend that the reader interested in probabilistic models of games should read Sections 1 and 2 of Chapter 2.

Since the problems we discuss have an immediate connection with programming, we have felt it worthwhile to introduce some notation taken from programming languages, in particular ALGOL. We use the assignment symbol ': =' and we denote the end of the definition of an operator by the semicolon ';'. Whenever this convention is in conflict with the ordinary rules of syntax we give precedence to the formal language.

We have devoted many years, and still devote our time, to the development of programs for playing chess. Quite naturally, this devotion has influenced our exposition; in particular, the examples we present are often related to chess. Nevertheless, with a few exceptions we have avoided the discussion of problems that arise uniquely in chess. Moreover, we are firmly convinced that one can write strong chess-playing programs without being a good chess player. Therefore, even when the examples presented deal purely with chess, the reader needs only an elementary knowledge of the game.

Translator's Note

The reader should be aware that the Russian text of this book was published in 1978, and that in the years since then, progress in the development of computer chess programs has been rapid. The book is still of interest since it deals with the ideas basic to the problems of search rather than with specific programs and programming techniques. Readers interested in recent developments should consult the publications of the International Computer Chess Association, (*Journal of the ICCA*, edited by H.J. van den Herik, Department of Informatics, University of Technology, Julianalaan 132, 2628 BL, Delft, The Netherlands). Other publications of interest include a recent paper by T. A. Marsland which cites the more important work in this field (*Computer Chess Methods*, *Encyclopedia of Artificial Intelligence*, Wiley, 1987, pp. 159–171), and Hartmut Tanke's very complete *Computer Chess Bibliography* (recently reissued in German). This bibliography can be obtained by writing to: Hartmut Tanke, Kienitzer Str. 104–106, D–1000 Berlin 44, West Germany.

CHAPTER 1

Two-Person Games With Complete Information and the Search of Positions

The Game Tree, Position Score, and Best Move

We recall the recursive definition of a tree. The *elements of a tree* are *nodes* and *arcs*; one of the incident nodes of an arc is called the *beginning* and the other the *end*.

A *tree* is either: a) a single node, or b) a tree with an additional node and an arc beginning at an already existing node and ending at the new node. If the old node is denoted by A and the new one by B, the arc leading from A to B is denoted by (A, B) or $\vec{B}$ *). It is easy to see that every tree is connected. (See Figure 1.)

The *base of a tree* is a node to which no arc leads. A *terminal* node is one from which no arc issues. It follows easily from the definition that every tree has a unique base.

A *subtree of a tree* $\mathfrak{A}$ is a subset of its nodes and arcs that is itself a tree.

Let $\mathfrak{B}$ and $\mathfrak{C}$ be subtrees of a tree $\mathfrak{A}$ with a non-empty intersection $\mathfrak{D}$. Then $\mathfrak{D}$ is a subtree of $\mathfrak{A}$.

A node B is said to be *subordinate* to A if

a) $B \equiv A$ or b) B is the end of an arc issuing from a node subordinate to A. An arc issuing from a node subordinate to A is said to be subordinate to A.

The set of arcs and nodes subordinate to a node A forms a tree with the base A. This tree is called the *A-subtree* of the original tree.

---

* In graph theory, one may assume that the arcs are undirected, i.e. the beginning and end are not distinguished. The object we have just defined is known as a directed graph. (The notions of tree and directed graph may also be defined in different ways.) However, since nowhere in the text do we discuss undirected graphs, the term *directed graph* is everywhere replaced by *tree*.

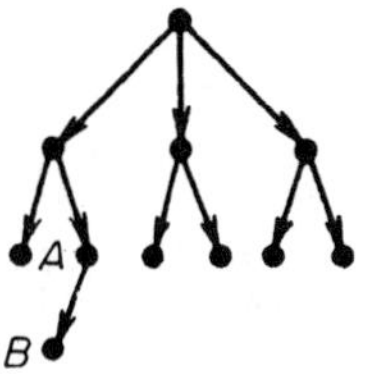

Figure 1

If A is not the base of the original tree, the A-subtree together with the arc $\vec{A}$ is called the *open* A-subtree.

If in a tree $\mathfrak{A}$ we omit the arc leading to a node A and also the A-subtree the remainder $\mathfrak{B}$ is a tree.

The base of a tree will be called a node of rank 0. If the node A is of rank n and there exists an arc (A, B) the node B has rank $n+1$. The *rank* of each node in a tree is defined in this way.

Let A be a node of a tree and let B be a node of the A-subtree. The *branch* $W(A, B)$ is a sequence of nodes $A_0 = A, A_1, A_2, \ldots, A_k = B$ such that the arcs (A_s, A_{s+1}) exist for all $s = 0, 1, \ldots, k-1$. If the branch $W(A, B)$ exists it is unique.

Suppose that a color—White or Black—is assigned to every node in a tree, and that real numbers $\mathrm{sc}(F)$, called *scores* are assigned to some of the terminal nodes F. Then we call the tree a *game tree for two opponents*, or a *two-person game tree*.

The nodes of such a tree are called *positions* (the terminal nodes are called *terminal* positions), and the arcs are called *moves*. The moves leading from a position colored White (Black) are called White moves (Black moves). The corresponding positions are called *positions with White* (*Black*) *to move*, or *White* (*Black*) *positions*. (A game tree may be infinite, and many of our subsequent assertions hold for infinite trees.)

Let $\mathfrak{A}$ be a game tree with White to move in the base position A_0. This means that White must choose a move from among all the possible moves originating at A_0. After this move a new position A_1 arises, at the end of the arc (A_0, A_1) of the game tree. If Black has the move at A_1 he must also choose one move from among those issuing from A_1, and so on. In this way, when White and Black have exercised all their successive choices we arrive at a sequence $A_0, A_1, \ldots, A_n$ in which the last term is a terminal position with the score $\mathrm{sc}(A_n)$. We shall call this sequence a *game* and the score of the terminal position the *outcome of the game*. White chooses his moves with the aim of maximizing the outcome, and Black chooses his with the aim of minimizing it. The *score of the base position* A_0 is the outcome that both White and Black succeed in obtaining. It is not immediately clear, however, that both White and Black should count on obtaining the same outcome.

Let $\mathfrak{A}$ be a game tree. We shall say that $\mathfrak{B}(A)$ is a *W-pruned* tree (relative to $\mathfrak{A}$) if

(1) $\mathfrak{B}$(A) is obtained from the A-subtree by excluding some number of open B-subtrees, where $\vec{B}$ is a White move.

(2) All terminal positions in $\mathfrak{B}(A)$ are terminal positions in $\mathfrak{A}$, i.e. no new positions arise as a result of the exclusion. A B-pruned subtree is similarly defined for Black.

We use the symbolic inequality

$$\mathrm{sc}(A) > M$$

when there exists a finite W-pruned tree with the base A for which all terminal positions have a score and all scores exceed M. (The inequality symbol may for the moment be regarded as an hieroglyph; we shall show later that in the contemplated game White may count on obtaining a numerical outcome $> M$.)

Similarly, if M exceeds all the terminal scores of some B-pruned tree with the base A we write

$$\mathrm{sc}(A) < M.$$

For the case of weak inequality the change in the definitions is obvious.

Let $\mathfrak{A}$ be a game tree and $\mathfrak{B}$ a W-pruned subtree of it. Then for every Black position $A \in \mathfrak{B}$, $\mathfrak{B}$ contains all the moves leading from A in $\mathfrak{A}$, provided that A is not a terminal Black position. A similar assertion holds with the colors interchanged.

Lemma. *If $\mathfrak{B}_1(A)$ is a W-pruned tree of the game tree $\mathfrak{A}$ and $\mathfrak{B}_2(A)$ is a B-pruned tree, then $\mathfrak{B} = \mathfrak{B}_1(A) \cap \mathfrak{B}_2(A)$ is a tree and all its terminal positions are terminal positions of $\mathfrak{A}$.*

Proof. $\mathfrak{B}$ is not empty since it contains the base A; it is therefore a tree. Let $B \in \mathfrak{B}$ be a non-terminal node of $\mathfrak{A}$ with White (Black) to move. Then all the moves from B are contained in $\mathfrak{B}_2(A)$ ($\mathfrak{B}_1(A)$), and at least one move is contained in $\mathfrak{B}_1(A)$ ($\mathfrak{B}_2(A)$), i.e. B is not a terminal position in $\mathfrak{B}$. □

Theorem on Consistency. *If $\mathfrak{A}$ is a game tree and $A \in \mathfrak{A}$ the symbolic inequalities*

$$\mathrm{sc}(A) > M,$$

$$\mathrm{sc}(A) \le M$$

cannot both hold.

Proof. The symbolic inequality $\mathrm{sc}(A) > M$ implies that there exists a W-pruned tree $\mathfrak{B}_1$ with base position A and scores for all its terminal positions $> M$; the inequality $\mathrm{sc}(A) \le M$ implies the existence of a B-pruned tree with the same base A and all scores of its terminal positions $\le M$. The intersection of these two trees contains at least one terminal position F with a score that both exceeds and does not exceed M. □

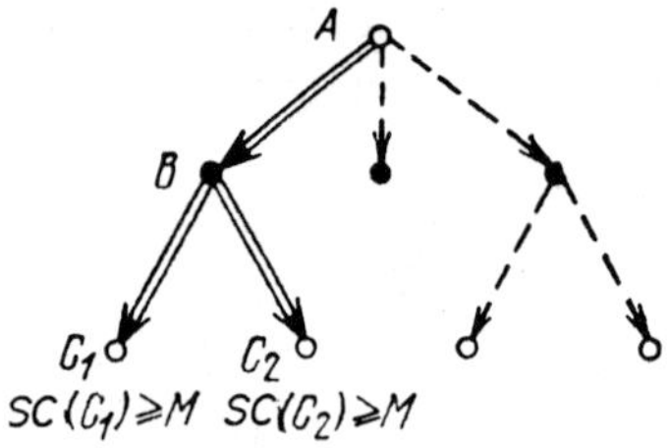

Figure 2

Definition. If we have both $\mathrm{sc}(A) \geq M$ and $\mathrm{sc}(A) \leq M$ for some position A in a game tree $\mathfrak{A}$, we say that M is the *score of the position A*.

From here on, all our arguments concerning positions, moves, and pruned trees will hold when colors and directions of inequalities are simultaneously interchanged. As a rule, we will make no specific mention of this fact.

Lemma. *Let A be a position in the game tree* $\mathfrak{A}$ *with White to move and let* $\vec{B}$ *be a move issuing from it such that* $\mathrm{sc}(B) \geq M$ $(> M)$. *Then* $\mathrm{sc}(A) \geq M$ $(> M)$.

Proof. Since $\mathrm{sc}(B) \geq M$ there exists a W-pruned tree $\mathfrak{B}$ with base B, for which all terminal positions have scores $\geq M$. Then $\mathfrak{B} + A$ (see Figure 2) is also a W-pruned subtree with the base A and the same terminal positions as $\mathfrak{B}$. □

Lemma. *Let A be a position in the game tree* $\mathfrak{A}$ *with White to move and let* $\vec{A}_1, \vec{A}_2, \ldots, \vec{A}_n$ *be the complete set of moves issuing from A. If the inequality* $\mathrm{sc}(A_s) \leq M$ $(< M)$ *holds for all the positions* A_s $(s = 1, 2, \ldots, n)$, *then* $\mathrm{sc}(A) \leq M$ $(< M)$.

Proof. Each of the inequalities $\mathrm{sc}(A_s) \leq M$ $(< M)$ follows from the existence of the B-pruned tree $\mathfrak{B}_s$ with base A_s and all scores of terminal positions $\leq M$ $(< M)$. Then $A + \mathfrak{B}_1 + \mathfrak{B}_2 + \cdots + \mathfrak{B}_n$ (see Figure 3) is a W-pruned tree with base A, with all its terminal positions $\leq M$ $(< M)$. □

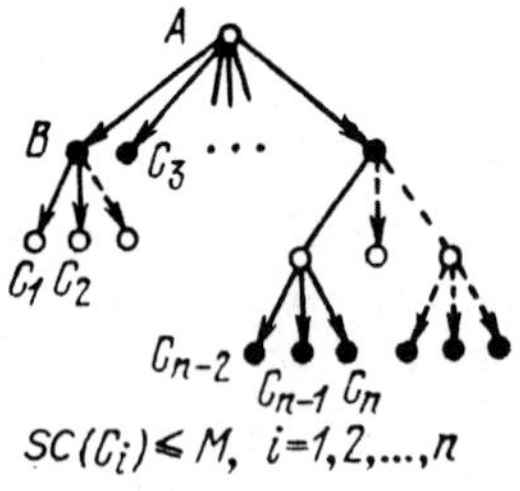

Figure 3

Theorem on the Transfer of Scores. *Let A be a node of the game tree* $\mathfrak{A}$ *with White (Black) to move, and let* $\vec{A}_1, \vec{A}_2, \ldots, \vec{A}_k$ *be all the moves issuing from it. If each node* A_s $(s=1,2,\ldots,k)$ *has the score* $\mathrm{sc}(A_s)$, *the node A has the score* $\mathrm{sc}(A)=\max\{\mathrm{sc}(A_s)|s=1,2,\ldots,k)$. *If Black has the move at A then* $\mathrm{sc}(A)=\min\{\mathrm{sc}(A_s)|s=1,2,\ldots,k)$.

Theorem on the Existence of Scores. *If* $\mathfrak{A}$ *is a finite game tree and all its terminal positions have a score, every position A in* $\mathfrak{A}$ *has a score, which is completely determined by the A-subtree.*

PROOF. All the positions of maximum rank in the A-subtree of the game tree $\mathfrak{A}$ have scores, since they are terminal positions. If all positions A with rank exceeding n have scores, the theorem on the transfer of scores implies that all positions of rank n also have scores. □

Thus the scores of the positions A in the game tree $\mathfrak{A}$ satisfy the conditions

$$\mathrm{sc}(A)=\max\{\mathrm{sc}(B)|B\in\mathfrak{B}(A)\} \text{ if } A \text{ is a White position,}$$
$$\mathrm{sc}(A)=\min\{\mathrm{sc}(B)|B\in\mathfrak{B}(A)\} \text{ if } A \text{ is a Black position,}$$

where $\mathfrak{B}(A)$ is the set of positions B that are immediate successors of A, i.e. those for which there exists an arc (A, B) issuing from A. We shall call these the Zermelo formulae, since they were established by him in [35] as a definition of the score of a non-terminal position.

John von Neumann [26] gave a different definition of the score. He based it on the notion of a strategy, i.e. an initially formulated choice of the move for each position in a game tree at which the move belongs to one's own color. If White chooses a strategy s from his set of strategies S_w, and Black chooses $s'\in S_b$, the game to be played is uniquely defined. This game normally leads to a terminal position F having the score $\mathrm{sc}(F)$. Then its *result* $r(s,s')$ is defined as $\mathrm{sc}(F)$.

von Neumann proposed that one should choose a strategy such that it yields the best result obtainable against the most damaging strategy that can be chosen by one's opponent. He assigned this result as the score of the base position. In this way he defined two scores for the base position A_0 in the game tree $\mathfrak{A}$:

$$\mathrm{sc}_w(A_0)=\max\{\min(r(s,s')|s'\in S_b)|s\in S_w\},$$
$$\mathrm{sc}_b(A_0)=\min\{\max(r(s,s')|s\in S_w)|s'\in S_b\}.$$

The score for an arbitrary position B is similarly defined.

If the game tree is finite and every terminal position has a score, then for an arbitrary position B we have

$$\mathrm{sc}_w(B)=\mathrm{sc}_b(B)=\mathrm{sc}(B)$$

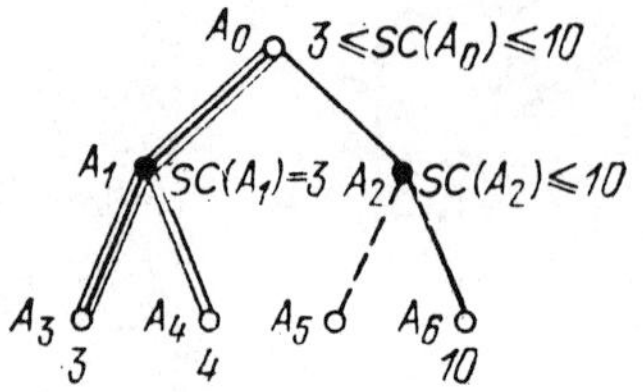

Figure 4

where $\mathrm{sc}(B)$ is the score of the position B according to our earlier definition. Thus if one of the players has chosen a strategy s that is optimal in the von Neumann sense, and his opponent has noted his choice, the opponent will know what answer will be made to each of his moves, yet will be unable to improve the result of the game for himself.

When scores are lacking at some of the terminal positions of the game tree $\mathfrak{A}$ with base A_0, there may nevertheless exist W-pruned and B-pruned trees with base A_0 and terminal positions that do have scores (see for example Figure 4). For these trees we will have the symbolic equations

$$\mathrm{sc}(A_0) \geq m,$$
$$\mathrm{sc}(A_0) \leq M,$$

and the consistency theorem implies that $m \leq M$. It is easy to find a game tree for which the score of the base position exists in our sense, but not in the von Neumann sense (see Figure 5). Since there exist strategies $s' \in S_w$ and $s'' \in S_b$ such that they prescribe the game (A_0, A_2, A_4), which has no result, both $\min\{r(s', s'')|s \in S_b\}$ and $\max\{r(s', s'')|s'' \in S_w\}$ are undefined, and so therefore are $\mathrm{sc}_w(A_0)$ and $\mathrm{sc}_b(A_0)$ undefined. At the same time there exist W-pruned and B-pruned trees with the base A_0 that define the value of $\mathrm{sc}(A_0)$ as 10.

Suppose given the game tree $\mathfrak{A}$, a subset $\mathfrak{F}$ of its terminal positions, and the scores of the positions $C \in \mathfrak{F}$. In some way or other we fix the values of the scores at terminal positions not contained in $\mathfrak{F}$. The resulting game tree $\mathfrak{A}'$ is called an *extension* of the game tree $\mathfrak{A}$ that we began with. Scores exist for the base position, and in general for all positions, of an arbitrary extension $\mathfrak{A}'$. It is therefore meaningful to speak of upper and lower bounds for the score $\mathrm{sc}(A_0)$ at the base position A_0 of the game $\mathfrak{A}$, taken with respect to all extensions of $\mathfrak{A}$.

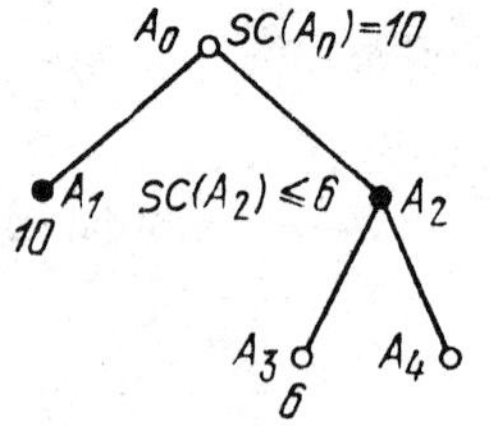

Figure 5

When we seek the score of the position A_0 of some game $\tilde{\mathfrak{A}}$ we may assume that the converse situation obtains. In order to avoid looking at the entire tree $\tilde{\mathfrak{A}}$, we try to find a tree $\mathfrak{A}$ for which our game is an extension, and for which

$$\sup\{\mathrm{sc}_{\mathfrak{A}'}(A_0)|\mathfrak{A}' \supset \mathfrak{A}\} = \inf\{\mathrm{sc}_{\mathfrak{A}'}(A_0)|\,\mathfrak{A}' \supset \mathfrak{A}\} = \mathrm{sc}_{\tilde{\mathfrak{A}}}(A_0).$$

Theorem on the Extensions of a Game. *The inequality*

$$\inf\{\mathrm{sc}(A_0)|\mathfrak{A}' \supset \mathfrak{A}\} > M \ (\geq M)$$

holds if and only if the symbolic inequality $\mathrm{sc}(A_0) > M \ (\geq M)$ *holds for the contemplated game* $\mathfrak{A}$, *i.e. there exists a W-pruned subtree* $\mathfrak{A}_w \subset \mathfrak{A}$ *for which all the terminal positions have scores* $> M \ (\geq M)$.

PROOF. A W-pruned subtree $\mathfrak{A}_w$ of the game $\mathfrak{A}$ is a W-pruned subtree of any extension $\mathfrak{A}'$ of $\mathfrak{A}$. Therefore if the scores of all terminal positions of $\mathfrak{A}_w > M \ (\geq M)$ we have for an arbitrary extension $\mathfrak{A}' \supset \mathfrak{A}$

$$\mathrm{sc}_{\mathfrak{A}'}(A_0) > M \ (\leq M)$$

and so

$$\inf\{\mathrm{sc}(A_0)|\mathfrak{A}' \supset \mathfrak{A}\} > M \ (\geq M).$$

On the other hand,

$$\inf\{\mathrm{sc}_{\mathfrak{A}'}(A_0)|\,\mathfrak{A}' \supset \mathfrak{A}\} = \mathrm{sc}_{\mathfrak{A}^-}(A_0),$$

where the terminal positions with undefined scores in $\mathfrak{A}$ have the score $-\infty$ in $\mathfrak{A}^-$ and then the inequality holds because $\mathrm{sc}(A_0)$ is a monotone function of the scores of the terminal positions in the game $\mathfrak{A}'$.

If $\inf\{\mathrm{sc}(A_0|\mathfrak{A}' \supset \mathfrak{A})\} = \mathrm{sc}_{\mathfrak{A}^-}(A_0) > M \ (\geq M)$, there exists a W-pruned subtree $\mathfrak{A}_w^-$ of the tree $\mathfrak{A}^-$ for which the scores of terminal positions $> M \ (\geq M)$. In this subtree all the terminal positions belong to the set $\mathfrak{F}$ (the scores of the remaining positions are set equal to $-\infty$) i.e. $\mathfrak{A}_w^-$ is a W-pruned subtree of the game $\mathfrak{A}$. □

The corresponding theorem on the upper bounds of scores of the base position A_0, taken with respect to all extensions of $\mathfrak{A}$, is formulated and proved in a similar fashion. These two theorems imply that the symbolic inequalities $\mathrm{sc}(A) > M \ (\geq M, \ < M, \ \leq M)$ may be understood as actual inequalities satisfied by the scores of a position A in an arbitrary extension $\mathfrak{A}'$ of a game $\mathfrak{A}$ (in the game $\mathfrak{A}'$ all terminal positions have scores, and therefore the score of an arbitrary position exists). If the score $\mathrm{sc}_{\mathfrak{A}}(A)$ exists, then A will have the same score in an arbitrary extension $\mathfrak{A}'$.

Clearly, there exist game trees for which there are no W-pruned and B-pruned subtrees bounding the score of the base position.

Suppose given the game tree $\mathfrak{A}$ with base position A_0 having the score $\mathrm{sc}(A_0)$. Then White's aim is to find the *best move* from the position A_0, i.e a move (A_0, B) leading to a position of rank 1 having the greatest of the possible scores. From what we have shown earlier, if White succeeds then $\text{sc}(A_0) = \text{sc}(B)$. The best move from a White position A other than the base position is similarly defined, since A is the base of a subtree with White to move.

Now suppose that A is a Black position. From Black's point of view, the best move (A, B) leads to a position having the least of all possible scores. The theorem on the transfer of scores implies that in this case, also, $\text{sc}(A) = \text{sc}(B)$.

Obviously, there may be more than one best move at any given position.

A branch $W(A, B)$ is said to be *critical* if each of its arcs represents a best move. It is easily seen that the scores of all positions on a critical branch have the same value. This property is diagnostic: if the scores at all positions on a branch $W(A, B)$ coincide, the branch is critical.

Let A be a White position. The move (A, B) will be a best move if and only if there exist a threshold value of M, a W-pruned tree $\mathfrak{A}_w(B)$ with base B for which the scores of all terminal positions are not less than M, and B-pruned trees $\mathfrak{A}_b(B')$ for all other moves (A, B'), in which the scores of all terminal positions do not exceed M. In this case, $\text{sc}(B) \geq M$, $\text{sc}(B') \leq M$ if $(A, B') \in \mathfrak{A}$ *and* $B' \neq B$. Therefore

$$\text{sc}(A) = \max\{\text{sc}(B') | (A, B') \in \mathfrak{A}\} = \text{sc}(B).$$

Conversely, if (A, B) is a best move, $\text{sc}(A) = \text{sc}(B)$ and $\text{sc}(B') \leq \text{sc}(A)$ if $(A, B') \in \mathfrak{A}$. We may take $\text{sc}(A)$ as the threshold value M. The conditions for a best move at a Black position are similarly formulated.

When the game tree is not too large and both players are able to investigate it, e.g. to construct W-pruned and B-pruned trees, the definition given above corresponds to the intuitive notion of a best move. Not so when at least one of the players cannot perceive all the possibilities. Let us look at the game depicted in Figure 6, in which White is aware of all the variations to a depth of four plies and Black sees to a depth of two plies only.

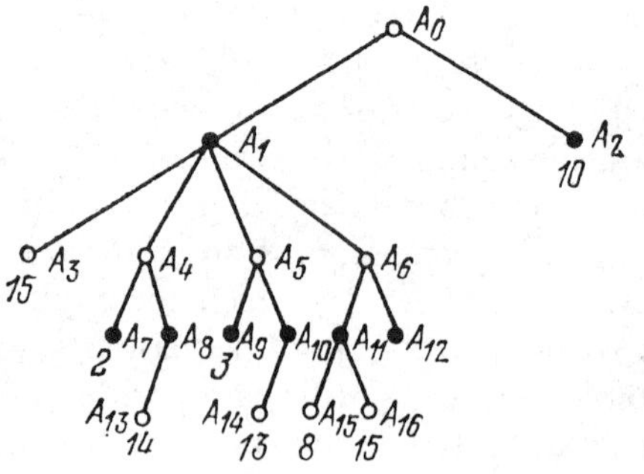

Figure 6

White can make the dismal move (A_0, A_2) leading to the position A_2 with a score of 10. The other move (A_0, A_1) offers a choice for Black: reply with the cautious move (A_1, A_3) and arrive at the position A_3 with a score of 15, or choose one of the three moves (A_1, A_4), (A_1, A_5), (A_1, A_6), for which he cannot see the consequences. His best move (A_1, A_6) leaves White with the score 8, so that the cautious move (A_0, A_2) is indeed best for White. However, we might suppose that Black would act differently. The move (A_1, A_6) is ostensibly no better than two others, which are in fact more enticing since Black perceives the replies (A_4, A_7) and (A_5, A_9) as more favorable for him. In every case where a choice of move must be made repeatedly in a similar situation, an active move such as (A_0, A_1) will lead to better results.

A player who cannot see the whole tree is in the same state as a player in a game with incomplete information. The information he has about the game can be represented by a tree in which some of the terminal positions have unknown scores. One might treat these scores as random variables, estimate in some way or other their stochastic parameters (mean, variance, correlation with scores in other positions) and try to choose a move that will maximize the expected outcome, or maximize the probability of choosing the best move (these objectives are not at all equivalent).

A player who sees to a greater depth than his opponent, and has a representation of the latter's probabilistic model, can build his own probabilistic model, which will enable him to choose risky moves that will increase the probability of winning.

We shall return to the problem of constructing various games that model some given game, but first we must consider some algorithms for finding the best moves in the sense of the formal definition given above.

Searching the Positions in a Game to Define the Score of the Base Position

All, or almost all, of the scores in a game tree $\mathfrak{A}$ can be determined by inspecting the positions in some order $L\{A_1, A_2, \ldots, A_m\}$, perhaps involving repetitions. The choice of successive elements of this sequence is a major part of the work to be done by any algorithm for computing scores. In almost all of the known algorithms the choice is made by traversing the nodes of the tree $\mathfrak{A}$, i.e. the list L satisfies the following condition: if the arc (B, C) belongs to the tree and both its nodes B and C are in L, then the position B is last encountered later than C, i.e. $\max\{i|A_i = B) > \max\{j|A_j = C)$.

If all the nodes of $\mathfrak{A}$ appear in L, the latter is called an *exhaustive search*. An exhaustive search allows the determination of the scores of all positions in the game, by application of Zermelo's formula whenever a non-terminal position is last encountered in it.

Many different methods are used in the algorithms for searching the positions in a tree $\mathfrak{A}$, and many of them will be considered in this book. First, however, we must devote some time to the widely used *backtracking* search. An exhaustive backtracking search of a tree $\mathfrak{A}$ is the traversal of a list $L\{A_0, A_1, \ldots, A_0\}$ in which the successive elements satisfy the following conditions:

(1) The first element is the base A_0 of the tree $\mathfrak{A}$.
(2) If the i-th element is the node B, and there exist arcs (B, C) whose ends have not yet been encountered in L, then the $(i+1)$-st element is the end of one of these arcs.
(3) Otherwise the $(i+1)$-st element is the beginning A of a unique arc (A, B) of $\mathfrak{A}$ with its end at the node B (if $B = A_0$ no such arc exists and the i-th element of the list $L(A_0, A_1, \ldots, A_0)$ is the last). In short, we go forward if we can and if we cannot we go backward.

A tree can be so represented that a forward step is always made to the leftmost available node. (See Figure 7.) In this representation nodes of the same rank are taken in order from left to right.

A backtracking search of the nodes of a tree is economical. The terminal nodes are visited only once each. Also, any arc $(B, C) \in \mathfrak{A}$ is traversed twice only—once in the forward direction and once backward—by a *forward step* from B to C and a *backward step* from C to B. Since to any pair A_i, A_{i+1} of neighboring elements in the search there correspond a forward step and a backward step, the number of steps is greater by one than double the number of arcs, or less by one than double the number of nodes (the number of nodes is greater by one than the number of arcs). Thus, on the average each node in the tree is encountered twice.

At every point in an exhaustive backtracking search the subset $\mathfrak{B}_p$ of nodes not yet encountered and the subset $\mathfrak{B}_F$ of nodes that will be encountered again are defined by information about the nodes in the branch $W(A_0, \ldots, A_i)$ connecting the base A_0 to the next element A_i of the sequence. It suffices to know, for each node $A_h \in W$, which of the ends B of the arcs (A_h, B) have already been met in the search. In fact, if the node B is not in W and has already been met in the search, all the nodes in the B-subtree will have been visited and will not be visited again. If B has not yet been visited, the nodes in its subtree will not have been visited either; only the nodes in the branch W have been visited and will be revisited. We have enough information for continuation of the search after such a node is met for the second time. Thus a backtracking search is easily carried out even when the tree is not given explicitly and we have only an algorithm for generating the arcs (A, B) issuing from a given node A, together with their ends B.

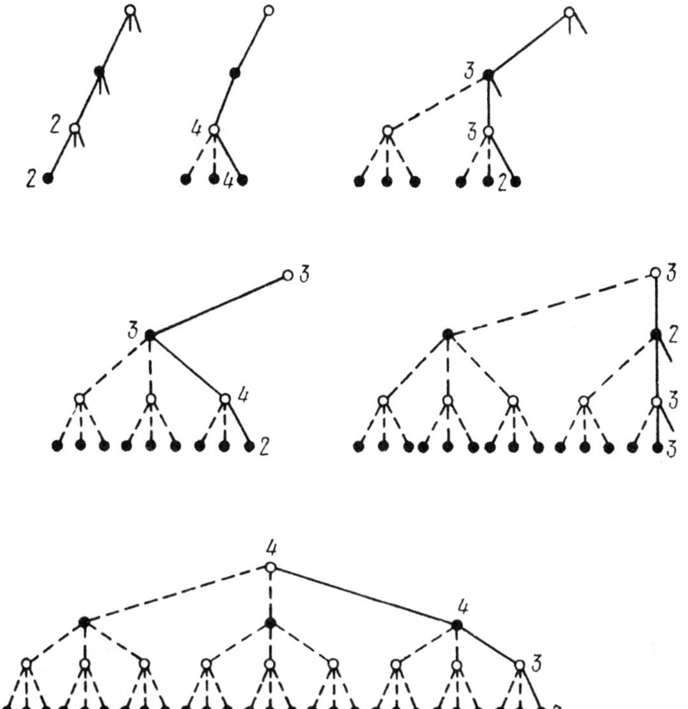

Figure 7

The scores of all the nodes in a game tree can be determined in parallel with the search by recording for all positions $A_h \in W$ the *partial score* $\mathrm{psc}(A_h)$, which has the following values:

$\mathrm{sc}(A_h)$ if A_h is a terminal position;
$\max\{\mathrm{sc}(B) | B \in \mathfrak{B}(A_h)\}$ if A_h is a non-terminal White position;
$\min\{\mathrm{sc}(B) | B \in \mathfrak{B}(A_h)\}$ if A_h is a non-terminal Black position.

Here $\mathfrak{B}(A_h) = \mathfrak{W}(A_h) \cap \mathfrak{B}_p$ is the set of all immediate successors of A_h that have already been investigated. For a non-terminal position with $\mathfrak{B}(A)$ empty we may write $\mathrm{psc}(A) = \pm\infty$ ($-$ for a White position, $+$ for a Black). Thus, if $\mathfrak{B}(A) = \mathfrak{W}(A)$, $\mathrm{psc}(A) = \mathrm{sc}(A)$.

To start the search we set $\mathrm{psc}(A_0) = \pm\infty$; after a forward step to the end of the branch W we add a new node B and define $\mathrm{psc}(B) := \mathrm{sc}(B)$ if B is terminal, $:= -\infty$ if B is a non-terminal White position, $:= +\infty$ if B is a non-terminal Black position. A backward step from a node B occurs when no forward step can be made. This happens whenever all the immediate successors of B have been investigated, and then $\mathrm{psc}(B) = \mathrm{sc}(B)$. We delete B from the branch W, and to its immediate predecessor A we assign the partial score $\operatorname{psc}(A) := \max\{\operatorname{psc}(A), \operatorname{sc}(B)\}$ if A is a White position and $:= \min\{\operatorname{psc}(A), \operatorname{sc}(B)\}$ if A is a Black position.

At any point in the search, the partial scores so computed for positions in the branch W will satisfy the definition of a partial score and will equal the score itself when the position is encountered for the last time.

Let us now see how to find the W-pruned subtree $\mathfrak{A}_w$ and the B-pruned subtree $\mathfrak{A}_b$ that prescribe the score we have found for the base position A_0. To do this we represent the game tree $\mathfrak{A}$ in such a way that at every position the forward steps are chosen in order from left to right, and we *color* the moves as White or Black by the following rules:

(1) If White (Black) has the move in the base position A_0 all moves leaving A_0 are colored Black (White) and the leftmost move (a best move) is also colored White (Black).
(2) If a move of some color leads to a position A with White (Black) to move, the leftmost best move from it is given the same color.
(3) If a move of the opposite color leads to a position A with White (Black) to move, all moves leading from it are also colored in the opposite color.

Thus some W-pruned subtree $\mathfrak{A}_w$ of $\mathfrak{A}$ is colored White, and some B-pruned subtree $\mathfrak{A}_b$ is colored Black; both are minimal with respect to inclusion and they intersect in the leftmost branch W issuing from the base position A_0; the moves in this branch, and only these, are colored in both Black and White. (See Figure 8.)

Lemma on a Branch of the Tree $\mathfrak{A}_w$. *Let $W(A = A_1, \ldots, A_k)$ be an arbitrary branch of the W-pruned tree $\mathfrak{A}_w$. Then* $\operatorname{sc}(A_1) \le \operatorname{sc}(A_2) \le \ldots \le \operatorname{sc}(A_k)$.

In fact if $A_i \in W$ is a White position ($i < k$) then (A_i, A_{i+1}) is a best move from it and $\operatorname{sc}(A_i) = \operatorname{sc}(A_{i+1})$; if, however, A_i is a Black position, then

$$\operatorname{sc}(A_i) = \min\{\operatorname{sc}(B) | B \in \mathfrak{B}(A_i)\} \le \operatorname{sc}(A_{i+1}).$$

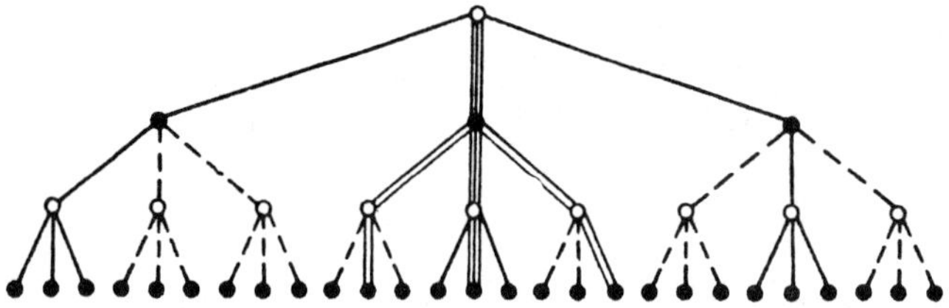

Figure 8

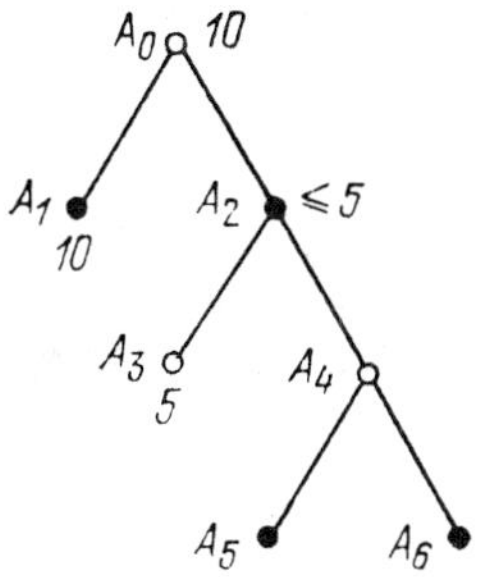

Figure 9

The lemma on branches of the Black-colored tree $\mathfrak{A}_b$ is formulated and proved in a similar way. These lemmas imply that the scores of all positions in the intersection $\mathfrak{A}_w \cap \mathfrak{A}_b$ are equal to sc(A_0). The move (A_0, A_1), which belongs to the critical branch $W = \mathfrak{A}_w \cap \mathfrak{A}_b$ is a best move; the scores of all terminal positions in the White tree $\mathfrak{A}_w$ are not less than sc(A_0) and the scores in the Black tree $\mathfrak{A}_b$ are not greater than sc(A_0). Therefore to determine the score of the position A_0 and find a best move from it we need consider only these subtrees, and any other work on the search is superfluous.

To find a best move from the base position A_0 we do not need to know the scores of all positions in the game tree $\mathfrak{A}$. Suppose for instance that $\mathfrak{A}$ is the game tree depicted in Figure 9, White and Black to move in alternating turns, with White to move in the base position A_0. Assume that we have carried out the portion $L'\{A_0, A_1, A_0, A_2, A_3, A_2\}$ of the search. (The scores of terminal positions and the partial scores of non-terminal positions are shown beside the corresponding nodes in Figure 9.) Since White is to move at A_0, $\mathrm{sc}(A_0) \geq \mathrm{psc}(A_0) = 10$; Black is to move at A_2 and $\mathrm{sc}(A_2) \leq \mathrm{psc}(A_2) = 5$. Moreover

$$\mathrm{sc}(A_0) = \max\{\mathrm{sc}(A_1), \mathrm{sc}(A_2)\} \leq \max\{\mathrm{sc}(A_1), \mathrm{psc}(A_2)\} = 10.$$

Accordingly $\mathrm{sc}(A_0) = 10$, (A_0, A_1) is a best move at the base position, and the search need not be extended. A more complicated example is shown in Figure 10, where White and Black alternate their moves, as in the preceding example. Suppose that we have carried out the portion $L'\{A_0, A_1, A_0, A_2, A_3, A_5, A_7, A_5\}$ of the search and have obtained partial scores for the positions A_0 and A_5. If (A_3, A_5) is a best move, then

$$\mathrm{sc}(A_2) \leq \mathrm{sc}(A_3) = \mathrm{sc}(A_5) \leq \mathrm{psc}(A_5) = 5.$$

Therefore we may skip the positions $A_8, A_9, \ldots, A_{13}$ and go at once to A_6. If it turns out that the score of A_6 is less than 10, we have

$$\mathrm{sc}(A_2) \leq \mathrm{sc}(A_3) = \max\{\mathrm{sc}(A_5), \mathrm{sc}(A_6)\} < 10.$$

i.e (A_0, A_1) is a best move at the base position A_0 and $\mathrm{sc}(A_0) = 10$. If $\mathrm{sc}(A_6) \geq 10$, $\mathrm{sc}(A_2) \geq \mathrm{sc}(A_3) = \mathrm{sc}(A_6) \geq 10$ and (A_0, A_2) is a best move.

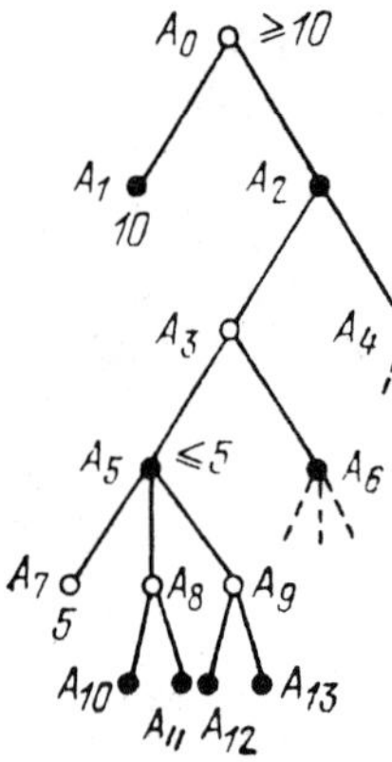

Figure 10

To shorten the search as much as we can we reformulate the second rule stated above to connect it with the partial scores that we are computing for the positions in the branch W:

2′. If the k-th element of the list L is a White (Black) position, and if for all Black (White) positions A in the branch $W(A_0, \ldots, B)$ we have $\mathrm{psc}(A) < \mathrm{psc}(B)$ ($\mathrm{psc}(A) > \mathrm{psc}(B)$), and if there exist arcs (B, C) with ends not yet encountered in the search, then the next element in the search is one of these new immediate successors of the position B.

We shall show later that we can in fact use this method to find a score for the base position A_0 and a best move from it. However, the values we compute, which we have called partial scores, are not really so. Rather, we shall call them *bounds*. The theory of bounds and scores was first considered by A. L. Brudno in his paper [10]. The consequent rule for cutting off a variation during the search, which is essentially equivalent to ours, is often called the *α,β-cutoff* or *branch-and-bound* heuristic. It is useful to reformulate it slightly in order to show more clearly just when the cutoff takes place, i.e. when a whole subtree of the game tree $\mathfrak{A}$ is thrown out of further consideration (pruned). The equivalence of the newly formulated rule with the old one will be obvious.

A *search* is a sequence of steps, to each of which is associated a next position B and either a next move (B, C) leading from B or a next move (A, B) leading to it. (We shall see later that there is one exception.) The next position for the following step is another position, related to the next move, and the next position for the first step is the base A_0. Thus at the beginning of every step we know the related next position. We do not know, however, whether the step is forward or backward (and even less do we know what the next move is). In order to formulate the rules for determining the character of the step we shall need the following notation: $\mathrm{bd}(A)$ is the value of the bound at the position A at the contemplated moment; $\mathrm{Bd}(A)$ is the last value of $\mathrm{bd}(A)$, i.e. the value at the instant of making a step backward from A; W is the branch leading from the base position to the next position; P_w is the set of White positions in W and P_b is the set of Black positions.

The Cutoff (Pruning) Rule: The next step is backward if the next position A is a White position and

$$\mathrm{bd}(B) \geq \min\{\mathrm{bd}(A) | A \in P_b\}$$

or if A is a Black position and

$$\mathrm{bd}(B) \leq \max\{\mathrm{bd}(A) | A \in P_w\}.$$

The Stopping Rule: The next step is backward if B is a terminal position or if all the immediate successors of B have already been examined.

The Rule for Deepening: If neither of the above rules is applicable, the next step is forward.

We now define the next move. If the step is backward it is the move (A, B) from A, the next preceding position but one in the branch W, to the next position B. If the next position is the base position no move is made; this signals the end of the search. If the step is forward the next move is (B, C) from the next position B to one of the positions not yet visited. We shall always suppose that we move to the leftmost of these. Thus after a backward step the branch W is shortened by one position and is lengthened by one position after a forward step.

Finally, we recompute, or determine, the value of the bound at the next position for the following step. If the step is backward, and the next move is (A, B) we have

$$\mathrm{bd}(A) := \max\{\mathrm{bd}(A), \mathrm{Bd}(B)\} \text{ if } A \text{ is a White position,}$$

$$\mathrm{bd}(A) := \min\{\mathrm{bd}(A), \mathrm{Bd}(B)\} \text{ if } A \text{ is a Black position.}$$

If the step is forward and the move is (B, C) we have

$$\mathrm{bd}(C) := \mathrm{sc}(C) \text{ if } C \text{ is terminal,}$$

$$\mathrm{bd}(C) := -\infty \text{ if } C \text{ is a non-terminal White position,}$$

$$\mathrm{bd}(C) := +\infty \text{ if } C \text{ is a non-terminal Black position.}$$

Before the beginning of the search the score of the base position A_0 is defined as if the first step were preceded by a blank move representing a forward step.

Thus at any point in the search the values of the bounds satisfy a condition like the Zermelo formula:

$\mathrm{Bd}(A) = \mathrm{sc}(A)$ if A is a terminal position; max $\{\mathrm{bd}(B) | B \in \mathfrak{B}(A)\}$ if A is a non-terminal White position; $\min\{\mathrm{bd}(B) | B \in \mathfrak{B}(A)\}$ if A is a non-terminal Black position. Here the maxima and minima are computed over only those immediate successor positions of A that have already been examined.

It is easily seen that the set of all positions examined in the search forms a subtree $\tilde{\mathfrak{A}}$ of the game tree $\mathfrak{A}$ with its root at the base position A_0 and terminal nodes corresponding to those positions of $\mathfrak{A}$ that have assigned scores. It may be looked on as the tree of a model game, with the same next move at each non-terminal position and the same scores at terminal positions as the original tree.

Let us now use the process described above to construct the White and Black subtrees $\tilde{\mathfrak{A}}_w$ and $\tilde{\mathfrak{A}}_b$ and show that these are respectively W-pruned and B-pruned subtrees of the original game tree $\mathfrak{A}$. Then $\mathrm{sc}(A_0) = \mathrm{bd}(A_0)$ and the best move is (A_0, A_1) where $A_1 \in \tilde{\mathfrak{A}}_w \cap \tilde{\mathfrak{A}}_b$. These assertions will be validated by three subsequent lemmas on the changes in the bounds during the search. We formulate and prove them with respect to White positions or positions in the tree $\tilde{\mathfrak{A}}_w$. The corresponding lemmas with an interchange of colors may be formulated and proved in a similar fashion. At any position $A \in \tilde{\mathfrak{A}}$ we write $\mathrm{bd}(A)$ for the next value of its bound, and $\mathrm{Bd}(A)$ for its final value.

Lemma on Monotone Change. *Let $A \in \tilde{\mathfrak{A}}$ be a White position. Then the value of* $\mathrm{bd}(A)$ *does not decrease.*

PROOF. The rules formulated above imply that $\mathrm{bd}(A)$ either takes on the value $\mathrm{sc}(A)$ immediately and remains constant, or takes on the value $-\infty$ and changes subsequently according to the rule

$$\mathrm{bd}(A) := \max\{\mathrm{bd}(A), \mathrm{bd}(B) | (A, B) \in \tilde{\mathfrak{A}}\},$$

i.e. can never decrease. □

Lemma on Strong Monotonicity. *Let $B \in \tilde{\mathfrak{A}}_w$ be a non-terminal White position. Until we step backward to it from its unique immediate next position in $\tilde{\mathfrak{A}}_w$, we have* $\mathrm{bd}(B) < \mathrm{Bd}(B)$.

PROOF. When we first arrived at B we had

$$\mathrm{bd}(B) := -\infty < \mathrm{Bd}(B).$$

If (B, C) is a move in a forward step from B to a position C in $\tilde{\mathfrak{A}}_w$ it is a best move in $\tilde{\mathfrak{A}}$, and no other forward step is a best move. Therefore, the final values of the bounds for positions to which these other moves lead will be less than $\mathrm{bd}(B) = \mathrm{bd}(C)$ and before a backward step from C to B we have

$$\mathrm{bd}(B) = \max\{\mathrm{Bd}(C') | C' \in \mathfrak{B}(B)\} < \mathrm{Bd}(B). \qquad \square$$

Lemma on the Completeness of the Subtree $\tilde{\mathfrak{A}}_w$. *Let $C \in \tilde{\mathfrak{A}}_w$ be a non-terminal White position. Then the cutoff rule is inapplicable at any position immediately preceding C.*

PROOF. Let $W(A_0, \ldots, C)$ be the branch from the base position A_0 to the Black next position C and let $B \in P_b$ *be any White position in the branch W. The final values of* $\mathrm{bd}(B)$ and $\mathrm{bd}(C)$ are equal to the scores of these positions in the game $\tilde{\mathfrak{A}}$. The B-subtree of the W-pruned tree $\tilde{\mathfrak{A}}_w$ is itself a

W-pruned tree with scores of its terminal positions not less than $\mathrm{bd}(B)$. It contains the position C and so also the C-subtree of $\tilde{\mathfrak{A}}_w$. This C-subtree is also a W-pruned tree. Therefore $\mathrm{Bd}(C) = \mathrm{sc}_{\tilde{\mathfrak{A}}}(C) \geq \min\{\mathrm{sc}(F) | F \in \tilde{\mathfrak{A}}_w \cap \mathfrak{F}\} \geq \min\{\mathrm{sc}(F) | F \in \tilde{\mathfrak{A}}_w(B) \cap \mathfrak{F}\} = \mathrm{sc}_{\tilde{\mathfrak{A}}}(B) = \mathrm{Bd}(B)$,where $\tilde{\mathfrak{A}}_w(B)$ and $\tilde{\mathfrak{A}}_w(C)$ are the B- and C- subtrees of $\tilde{\mathfrak{A}}_w$ and $\mathfrak{F}$ is the set of all terminal positions in the game $\mathfrak{A}$.

By the inequality proved above, and also by the lemmas on strong monotonicity and on the monotone change in the bound (we need the latter in the formulation for Black positions), the inequalities

$$\mathrm{bd}(B) < \mathrm{Bd}(B) \leq \mathrm{bd}(C) \leq \mathrm{Bd}(C)$$

hold as long as we have not made a backward step to the position B and so whenever we are considering the position C. Accordingly, when C is the next position we have for all $B \in P_w$ the inequality $\mathrm{bd}(B) < \mathrm{bd}(C)$ and the pruning rule cannot be applied. □

So for every non-terminal Black position $C \in \tilde{\mathfrak{A}}_w$ all the moves from it also belong to $\tilde{\mathfrak{A}}_w$, and this proves the

Theorem on the White Tree of a Search with Pruning. *The White tree $\tilde{\mathfrak{A}}_w$ is a W-pruned subtree of the game tree $\mathfrak{A}$.*

The theorems on the White and Black search trees imply the theorem on the bounds of the base position and the critical branch. After the end of the search $\mathrm{Bd}(\mathrm{A}_0) = \mathrm{sc}(\mathrm{A}_0)$ and the intersection of the White and Black trees is the critical branch.

Let us now see what information we have about the position B at the instant when we take a backward step from it. Let $W(A_0, \ldots, B)$ be a branch leading from the base position A_0 to B; let P_w and P_b be the sets of White and Black positions in this branch, not counting B itself, and write

$$\underline{\lim} := \max\{\mathrm{bd}(A) | A \in P_w\},$$
$$\overline{\lim} := \min\{\mathrm{bd}(A) | A \in P_b\}.$$

where the bounds for the position are evaluated at the instant when a backward step is made from B. They remain unchanged, of course, during the whole time between a forward step to B and the subsequent backward step from B. If $\mathrm{P}_w = \varnothing$, $\underline{\lim} = -\infty$; if $\mathrm{P}_b = \varnothing$, $\overline{\lim} = +\infty$.

Lemma. $\overline{\lim} > \underline{\lim}$.

PROOF. We may restrict ourselves to the case in which $\overline{\lim} < +\infty$ and $\underline{\lim} > -\infty$. Suppose that

$$\overline{\lim} \leq \underline{\lim}, \qquad \underline{\lim} = \mathrm{bd}(A') | A' \in P_w,$$
$$\overline{\lim} = \mathrm{bd}(A'') | A'' \in P_b.$$

Clearly $A' \neq A''$ since at A' White has the move and at A'' the move is Black's. Suppose for instance that these positions occur in the branch W in the order $W(A_0,\ldots, A',\ldots, A'',\ldots, B)$. Then A'' belongs to the A'-subtree of $\mathfrak{A}$ and B belongs to its A''-subtree, and the search will have been carried out in the following order: First the value of $\mathrm{bd}(A')$ is established as equal to $\underline{\mathrm{lim}}$; then we step forward along the branch W from A', and after a while (perhaps at once) we step forward to A''. After this step the value of $\mathrm{bd}(A'')$ remains fixed at $\overline{\mathrm{lim}}$ and we step forward from A'' along the branch W in the direction of B, which ultimately becomes the next position. However, once we have made the backward step to the position A'' the value of $\mathrm{bd}(A'')$ is equal to $\overline{\mathrm{lim}}$ and we have the equation

$$\mathrm{bd}(A'') = \overline{\mathrm{lim}} \leq \underline{\mathrm{lim}} = \mathrm{bd}(A').$$

i.e., we must apply the pruning rule and therefore step backward from the position A''. But then the position B is never reached in the search, and this contradicts the hypothesis of the lemma. □

If $\mathrm{Bd}(\mathrm{B}) \leq \underline{\mathrm{lim}}$ the position *B is unreachable by White* (when B is a Black position the pruning rule is applied) and if $\mathrm{Bd}(\mathrm{B}) \geq \overline{\mathrm{lim}}$, B is *unreachable by Black*. What this means will be explained later.

Theorem on Bounds and Scores. *If for a backward step from a position B we have* $\mathrm{Bd}(B) \geq \overline{\mathrm{lim}}$ *then* $\mathrm{sc}(B) \geq \overline{\mathrm{lim}}$; *if* $\mathrm{Bd}(B) \leq \underline{\mathrm{lim}}$ *then* $\mathrm{sc}(B) \leq \underline{\mathrm{lim}}$; *if however* $\underline{\mathrm{lim}} < \mathrm{Bd}(B) < \overline{\mathrm{lim}}$, *then* $\mathrm{Bd}(B) = \mathrm{sc}(B)$.

PROOF. We consider an auxiliary game $\tilde{\mathfrak{A}}(B)$ for which the tree consists of the positions $\tilde{A}_0, \tilde{A}_1, \tilde{A}_2, \tilde{A}_3$, and the moves $(\tilde{A}_0, \tilde{A}_1)$, $(\tilde{A}_0, \tilde{A}_2)$, $(\tilde{A}_2, \tilde{A}_3)$, $(\tilde{A}_2, B)$, plus the B-subtree of the game $\mathfrak{A}$ (see Figure 11). The move at $\tilde{A}_0$ belongs to the side that has the move at B, and belongs to the opposite side at $\tilde{A}_2$ (we assume that B is a White position). The positions $\tilde{A}_1$ and $\tilde{A}_3$ are terminal and have the scores $\mathrm{sc}(\tilde{A}_1) = \underline{\mathrm{lim}}$, $\mathrm{sc}(\tilde{A}_3) = \overline{\mathrm{lim}}$ (if B is a Black position the converse holds, i.e. $\mathrm{sc}(\tilde{A}_1) = \overline{\mathrm{lim}}$, $\mathrm{sc}(\tilde{A}_3) = \underline{\mathrm{lim}}$). The turn to move at non-terminal positions in the B-subtree and the scores of terminal positions are the same as in the game $\mathfrak{A}$. Then that portion of the search of the tree $\mathfrak{A}$ that begins after the step forward from B and ends with a step backward from it is identical with the search of the game $\tilde{\mathfrak{A}}(B)$ after the initial steps with the next positions $\tilde{A}_0, \tilde{A}_1, \tilde{A}_0, \tilde{A}_2, \tilde{A}_3, \tilde{A}_2, B$, and after completion of this segment the search of the tree $\tilde{\mathfrak{A}}(B)$ ends in two backward steps with the next positions $\tilde{A}_2$ and $\tilde{A}_0$.

The theorem on the bounds of the base position implies that $\mathrm{Bd}(\tilde{A}_0) = \mathrm{sc}(\tilde{A}_0)$—remembering that $\mathrm{Bd}(\tilde{A}_0)$ is the value of $\mathrm{bd}(A_0)$ at the end of the search algorithm. Moreover,

$$\mathrm{sc}(\tilde{A}_0) = \max\{\underline{\mathrm{lim}}, \mathrm{sc}(\tilde{A}_2)\},$$

$$\mathrm{Bd}(\tilde{A}_0) = \max\{\underline{\mathrm{lim}}, \mathrm{Bd}(\tilde{A}_2)\}.$$

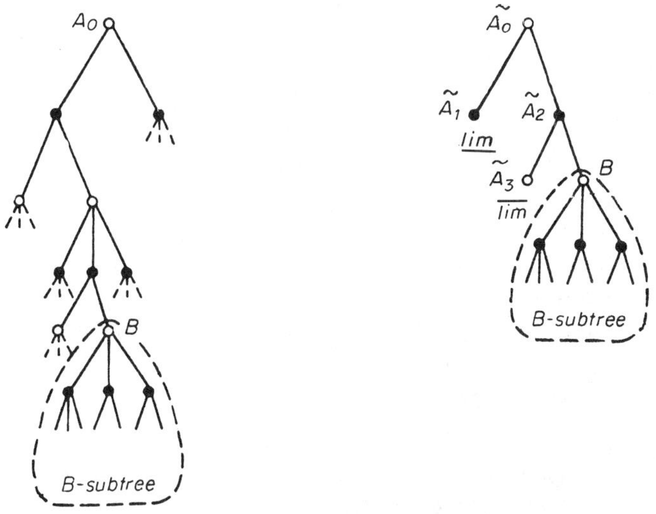

Figure 11

So, either $\operatorname{sc}(\tilde{A}_0)=\underline{\lim}$ and then

$$\operatorname{sc}(\tilde{A}_2)\le\underline{\lim},\qquad \operatorname{Bd}(\tilde{A}_2)\le\underline{\lim},\qquad \text{or}$$
$$\operatorname{sc}(\tilde{A}_0)=\operatorname{sc}(\tilde{A}_2)=\operatorname{Bd}(\tilde{A}_2)>\underline{\lim}.$$

In the first case $(\tilde{A}_2, B)$ is a best move from the position $\tilde{A}_2$, and another move leads to a position $\tilde{A}_3$ with the score $\overline{\lim}>\underline{\lim}\ge\operatorname{sc}(\tilde{A}_2)$. Accordingly

$$\underline{\lim}\ge\operatorname{sc}(\tilde{A}_2)=\operatorname{sc}(B),$$
$$\underline{\lim}\ge\operatorname{Bd}(\tilde{A}_2)=\min\{\overline{\lim},\operatorname{Bd}(B)\}=\operatorname{Bd}(B).$$

In the second case either

$$\operatorname{sc}(\tilde{A}_2)=\min\{\overline{\lim},\operatorname{sc}(B)\}=\overline{\lim}\le\operatorname{sc}(B),$$
$$\overline{\lim}=\operatorname{sc}(\tilde{A}_2)=\operatorname{Bd}(\tilde{A}_2)=\min\{\overline{\lim},\operatorname{Bd}(B)\}\le\operatorname{Bd}(B).$$

or

$$\overline{\lim}>\operatorname{sc}(B)=\operatorname{sc}(\tilde{A}_2)=\operatorname{Bd}(\tilde{A}_2)=\min\{\overline{\lim},\operatorname{Bd}(B)\}=\operatorname{Bd}(B).$$

Zermelo's formula implies that the position B has the same score in both of the games $\mathfrak{A}$ and $\tilde{\mathfrak{A}}(B)$. This concludes the proof of the theorem. A similar proof holds when B is a Black position. □

Thus a position B is unreachable by White if its score does not exceed $\underline{\lim}$, and is unreachable by Black if its score is not less than $\overline{\lim}$.

Theorem on Candidate Positions for the Critical Branch. *Unless a position is reachable by both White and Black it cannot lie on the critical branch defined by the search process.*

PROOF. Suppose for instance that

$$\mathrm{Bd}(B) \geq \overline{\lim} = \mathrm{Bd}(A) < +\infty, \; A \in P_b,$$

i.e. B is unreachable by Black. If it belonged to the critical branch $W(A_0, \ldots, A, \ldots, B)$, the latest values of $\mathrm{bd}(A)$ and $\mathrm{bd}(B)$ would be equal. But after a step backward from B the value of $\mathrm{bd}(B)$ does not change and the value of $\mathrm{bd}(A)$ cannot increase, since A is a Black position. Therefore $\mathrm{bd}(A)$ is already equal to $\mathrm{Bd}(B)$ and cannot change in the future. Meanwhile

$$\mathrm{bd}(A) = \min\{\mathrm{Bd}(B') | B' \in \mathfrak{B}(A)\} = \{\mathrm{Bd}(B'') | B'' \in \mathfrak{B}(A)\}.$$

Therefore the branch $W(A_0, \ldots, A, B'', \ldots)$, consisting of the positions examined during the search is also the critical branch and lies to the left of the branch W; that is, the latter is not constructed during the search, which is what we were to prove.

We treat the case $\mathrm{bd}(\mathrm{B}) \leq \underline{\lim}$ in the same way. □

Suppose we are interested only in knowing whether the condition $\mathrm{sc}(A) > m$ is satisfied at some position A in the game $\mathfrak{A}$, while we do not need an exact value of the score and are not interested in knowing the best move from A. Then we may search the A-subtree of the game, after setting the initial value of the bounds at non-terminal White positions to m rather than $-\infty$. We carry out the search as though at each White position B that we contemplate there existed a move (B, B') to a terminal position with $\mathrm{sc}(B) = m$, and this move is to be examined first. Then if $\mathrm{Bd}(A) > m$ it coincides with $\mathrm{sc}(A)$ and we can determine the best move. If, however, $\mathrm{Bd}(A) = m$ we have $\mathrm{sc}(A) \leq m$ and the best move is unknown. In the same way, when we wish to know whether $m < \mathrm{sc}(A) < M$ we write, at each forward step to a position B, $\mathrm{Bd}(B) := \mathrm{sc}(B)$ for a terminal position, $\mathrm{Bd}(B) := m$ for a non-terminal White position, and $\mathrm{Bd}(B) := M$ for a non-terminal Black position.

On the Number of Positions Examined in Determining the Score of a Position and Finding the Best Move

A game $\mathfrak{A}$ need not be defined by specifying its game tree and the scores of terminal positions. For instance, a chess position is defined by (a) the distribution of the pieces on the board, (b) the assignment of the turn to move, and (c) a small number of supplementary data (the right to castle on either side, capture en passant, drawn game after the third repetition of a position or by application of the fifty-move rule; these relate to the branch W connecting the given position to the base chess position). This information determines uniquely whether there is a succeeding terminal position and if so what its score is, or if there is no such position then what moves may be made. These specific details allow us, in many positions, to determine the score or the best move without pursuing the subordinate subtree all the way to its terminal positions.

If however the game tree is specified in some suitable (possibly indirect) way while the scores of the terminal positions are given explicitly and perhaps arbitrarily, the theorems on extensions of games for the definition of scores and best moves force us to consider all the terminal positions of certain W-pruned and B-pruned subtrees $\mathfrak{A}_w(A)$ and $\mathfrak{A}_b(A)$. On the other hand, the construction of these trees, and of the scores of terminal positions that satisfy the condition

$$\min\{\mathrm{sc}(C)|C \in \mathfrak{F}_w(A)\} = \max\{\mathrm{sc}(C)|C \in \mathfrak{F}_b(A)\}$$

suffices to define the score at A and the best move from A. Here $\mathfrak{F}_w(A)$ and $\mathfrak{F}_b(A)$ are respectively the sets of terminal positions of the subtrees $\mathfrak{A}_w(A)$ and $\mathfrak{A}_b(A)$.

If, as is usually the case, the W-pruned and B-pruned subtrees of the game $\mathfrak{A}$ are strongly branched because as a rule we can choose more than one move from any one of their non-terminal positions, the total number of positions in these subtrees is of the same order of magnitude as the number of their terminal positions. Therefore if we are looking for the most economical algorithm (with respect to number of positions examined) for finding position scores and best moves, we may look for it among those algorithms that examine the entire branch $W(A,\ldots,C)$ leading from the position of interest A to a terminal position C whenever C is under consideration in the process of finding the score and best move at A. We may suppose that A is the base position A_0, else we examine the game $\mathfrak{A}(A)$ whose tree is the A-subtree of the original game $\mathfrak{A}$, with the initial move belonging to the side that originally had it.

The branches $W(A_0,\ldots,C)$ define a subtree $\mathfrak{A}' \subset \mathfrak{A}$ which, by what we have said above, contains the W-pruned and B-pruned subtrees $\mathfrak{A}_w$ and $\mathfrak{A}_b$ with, respectively, the lower and upper bounds of the scores at the terminal positions. Since the game tree $\mathfrak{A}$ is finite these subtrees contain minimally inclusive subtrees $\mathfrak{A}'_w$ and $\mathfrak{A}'_b$ with the same property. We naturally expect the algorithm that examines the least number of position to be found among those that consider only the positions in the smallest W-pruned and B-pruned subtrees defining the score at the base position A_0.

Lemma on Minimality. *The W-pruned subtree $\mathfrak{A}_w$ of the game $\mathfrak{A}$ which is minimal inclusive contains precisely one move from every non-terminal White position B.*

In fact if $(B,C)\in\mathfrak{A}_w$ and $(B,C')\in\mathfrak{A}_w$, the second move and the C'-subtree may be excluded while the tree is still W-pruned. □

The lemma obtained by changing color also holds.

Lemma on the Critical Branch. *The minimal-inclusive W-pruned and B-pruned subtrees of a game tree $\mathfrak{A}$ intersect in a branch $W(A_0,\ldots,C)$ containing the base position A_0 and some terminal position C. If these subtrees determine the score at A_0, W is the critical branch.*

PROOF. Suppose that the non-terminal position $B\in\mathfrak{A}_w\cap\mathfrak{A}_b$, where $\mathfrak{A}_w$ and $\mathfrak{A}_b$ are the smallest W-pruned and B-pruned subtrees respectively. If B is a White (Black) position, then $\mathfrak{A}_w$ ($\mathfrak{A}_b$) contains exactly one move (B,C) leading from it and $\mathfrak{A}_w$ ($\mathfrak{A}_b$) contains all such moves. Thus $\mathfrak{A}_w\cap\mathfrak{A}_b$ contains a single move (A_0, A_1) leading from the base position A_0 to some first-rank position A_1, a single move from A_1 to a second-rank position A_2, and so on, until we arrive at the terminal position C. The intersection of two subtrees of the game tree $\mathfrak{A}$ is a tree, i.e. it is connected, and therefore the intersection $\mathfrak{A}_w\cap\mathfrak{A}_b$ contains no positions nor moves other than those making up the branch $W(A_0,\ldots,C)$. If the subtrees $\mathfrak{A}_w$ and $\mathfrak{A}_b$ determine the score at A_0, then for any position $B\in\mathfrak{A}_w$ we have

$$\mathrm{sc}(B)\geq\mathrm{sc}(A_0)$$

and for any position $B\in\mathfrak{A}_b$ we have

$$\mathrm{sc}(B)\leq\mathrm{sc}(A_0).$$

So, all the positions B in the branch $W(A_0,\ldots,C)$ have scores equal to $\mathrm{sc}(A_0)$, i.e. W is the critical branch, which is what we were to prove. □

Let $\mathfrak{A}_w$ and $\mathfrak{A}_b$ be W-pruned and B-pruned subtrees determining the score of the base position A_0 in the game $\mathfrak{A}$. Then there exist ways of choosing the next move in a search algorithm for the tree $\mathfrak{A}$ such that the algorithm leads only to nodes in the union $\mathfrak{A}_w\cup\mathfrak{A}_b$ of those subtrees. If $\mathfrak{A}_w$ and $\mathfrak{A}_b$ are W-pruned and B-pruned subtrees with the smallest number of nodes in their union, the contemplated variant of our algorithm is minimal with respect to all algorithms determining the score at the base position or the best move from it (without making use of any specific information about the scores at terminal positions). In choosing the next move, however, we assume that the trees $\mathfrak{A}_w$ and $\mathfrak{A}_b$ are known. Nevertheless, the proof we shall give below shows that if we are successful enough in guessing the best move when we are choosing the next moves in our forward steps, the number of positions examined in our search is more or less minimal.

The rule for choosing the next move is conveniently expressed, in the representation of the game tree $\mathfrak{A}$ on a plane surface, as the requirement that we choose the next move as the leftmost of all those possible. To

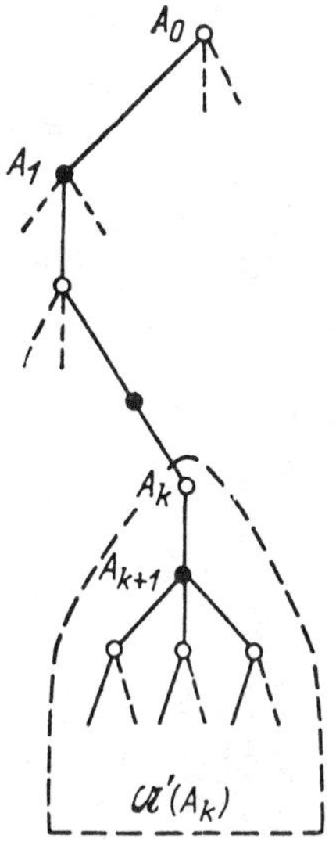

Figure 12

confine the search to the nodes in the union $\mathfrak{A}_w \cup \mathfrak{A}_b$ we must make the leftmost move $(B,C) \in \mathfrak{A}_w$ at every White node $B \in \mathfrak{A}_w$, and make the move $(B,C) \in \mathfrak{A}_b$ at every Black node $B \in \mathfrak{A}_b$. The rule for choosing the move at other nodes in the game tree $\mathfrak{A}$ is arbitrary. So when B lies on the critical branch $W = \mathfrak{A}_w \cap \mathfrak{A}_b$ we first attend to a move on this branch. Therefore at the beginning of the search we move through the nodes $A_0, A_1, \ldots, A_m$ of the branch W and we obtain the scores

$$\mathrm{Bd}(A_m) = \mathrm{sc}(A_m) = \mathrm{sc}(A_n),$$

and then from time to time we come back along W.

Lemma on the Bounds in the Critical Branch. *After a backward step to the position* $A_k \in W$, *we have* $\mathrm{Bd}(A_k) = \mathrm{sc}(A_k) = \mathrm{sc}(A_0)$.

PROOF. In fact the portion of the search that begins with the first forward step from the position A_k and ends with the first backward step to it may be taken as a search for the purpose of determining the score at the base position A_k of the game $\mathfrak{A}'(A_k)$, whose tree is the open A_{k+1}-subtree of $\mathfrak{A}$, and where the move $(A_k, A_{k+1}) \in W$. (See Figure 12.) Accordingly, after the first backward step to the position A_k, the theorem on bounds implies the equations

$$\begin{aligned}\mathrm{bd}(A_k) &= \mathrm{sc}_{\mathfrak{A}'(A_k)}(A_k) = \mathrm{sc}_{\mathfrak{A}'(A_k)}(A_{k+1}) \\ &= \mathrm{sc}(A_{k+1}) = \mathrm{sc}(A_0) = \mathrm{sc}(A_k),\end{aligned}$$

which is what we were to prove. □

Theorem on the Optimality of the Search. *Let $A_k \in W$ be a White position. Then the set of positions and moves that will be examined in the search, from the first step backward to A_k until the step backward from it, belongs to the subtree $\mathfrak{A}_b$.*

Proof. It suffices to show that the set of contemplated positions and moves that are subordinate to the immediate successor nodes B of A_k and that do not lie on the branch W do lie in the subtree $\mathfrak{A}_b$. Since $A_k \in \mathfrak{A}_b$ and is a White position, $B \in \mathfrak{A}_b$. Let us consider the portion of the search that begins after a forward step with the next move (A_k, B), and see when we next arrive at a position not lying in $\mathfrak{A}_b$. If we are at some position $D \in \mathfrak{A}_b$ and step backward from it, we land on an immediate predecessor position C that also belongs to $\mathfrak{A}_b$, since (C, D) is the only move in $\mathfrak{A}$ leading to D. If we step forward from $D \in \mathfrak{A}_b$ we again arrive at a position $E \in \mathfrak{A}_b$ since the move (D, E) is the leftmost. Finally, if we take any step forward from a White position we arrive at a node $E \in \mathfrak{A}_b$ since all moves from D lead to a position in $\mathfrak{A}_b$.

There remains the consideration of a backward step from a node $D \in \mathfrak{A}_b$ subordinate to B (and perhaps coinciding with it). We set the value of the bound at D as equal to the score at one of the terminal positions to be inspected after the first forward step from D. If the first forward step from the node B does not bring us to the end of the subtree $\mathfrak{A}_b$, none of the scores of the contemplated positions exceed $\mathrm{sc}(A_0)$. Thus $\mathrm{bd}(D) \leq \mathrm{sc}(A_0)$. But, by the lemma on the bounds in the critical branch, $\mathrm{bd}(A_k) = \mathrm{sc}(A_0)$, and we must step backward on account of the pruning rule, since A_k is a position in the branch $W'(A_0, \ldots, D)$ leading from the base position A_0 to the next position D.

Thus until we return to the position A_k and even more, until we take a backward step from A_k—the search moves only within the minimally inclusive B-pruned subtree $\mathfrak{A}_b$ of the game tree $\mathfrak{A}$. This is what we were to prove. □

There is a similar proof that after a first backward step to a Black position in the critical branch W and before any backward step from this position we contemplate only positions lying in the minimally inclusive W-pruned subtree $\mathfrak{A}_w$. Thus until we are forced to make a backward step from the base position A_0 we do not go outside the union $\mathfrak{A}_w \cup \mathfrak{A}_b$.

Let us now see how many terminal positions we need to inspect for the determination of the score at the base position when the moves alternate between the colors, i.e. when a White move (B, C) necessarily leads to a Black position, and vice versa. Suppose for instance that the base position, the only one of rank 0, is a White position. Then all positions of rank 1, and in general all positions of odd rank, are Black positions. The score at the base position can be obtained by a search employing the minimum amount of information about the scores at terminal positions.

We shall make a few remarks about the process. The backward steps and their next moves may be divided into three classes. If the value of $\mathrm{bd}(B)$ changes after a backward step with the next move (B, C) {in the way desired by the side that has the move at B} and begins or continues to satisfy the inequalities

$$\underline{\lim} < \mathrm{bd}(B) < \overline{\lim},$$

such a step and its next move (B, C) are said to be an *improving* step and move. After an improving step the pruning rule does not apply.

If $\mathrm{bd}(B)$ changes so that the pruning rule must be applied on the next step, the given step and its next move (B, C) are called a *refutation* step and move (one of the opponent's preferred moves in the branch). After a refutation step the values of $\mathrm{Bd}(B)$ and $\mathrm{Bd}(C)$ satisfy the conditions $\mathrm{Bd}(B) = \mathrm{Bd}(C) \geq \overline{\lim}$ if B is a White position, $\mathrm{Bd}(B) = \mathrm{Bd}(C) \leq \underline{\lim}$ if B is a Black position.

In the remaining cases the backward step and its next move are said to be *bad*, although in fact one of the moves in the branch preferred by the side of the same color may be bad. After a step backward from the position C we have $\mathrm{Bd}(C) \leq \mathrm{bd}(B) < \overline{\lim}$ if B is a White position, and $\mathrm{Bd}(C) \geq \mathrm{Bd}(B) > \underline{\lim}$ if B is a Black position. As in an improving step, so here the pruning rule does not apply to the following step. We will present some further discussion of White and Black positions, noting that in what we say we may simultaneously interchange colors, $\underline{\lim}$ and $\overline{\lim}$, and the inequality signs < and >, etc. Let (B, C) be the next move for some step in the search. The branch $W'(A_0, \ldots, B, C)$ leading from the base position A_0 to the position C is an extension of the branch $W(A_0, \ldots, B)$ from A_0 to B. Then the sets of positions in these branches with moves of one color that determine the values of $\underline{\lim}(B)$, $\overline{\lim}(B)$, $\underline{\lim}(C)$, $\overline{\lim}(C)$ satisfy the conditions

$$P_w(C) = P_w(B) \cup B,$$
$$P_b(C) = P_b(B).$$

If (B, C) is an improving move the inequality $\mathrm{Bd}(C) > \mathrm{bd}(B)$ holds before a backward step from C and the inequalities $\underline{\lim}(B) < \mathrm{bd}(B) = \mathrm{Bd}(C) < \overline{\lim}(B)$ hold after it. Thus, before such a step we have

$$\begin{aligned} \overline{\lim}(C) &= \min\{\mathrm{bd}(A) | A \in P_b(C) = P_b(B)\} \\ &= \overline{\lim}(B) > \mathrm{Bd}(C) > \max\{\mathrm{bd}(B), \underline{\lim}(B)\} \\ &= \max\{\mathrm{bd}(B), \max\{\mathrm{bd}(A) | A \in P_w(B)\}\} \\ &= \max\{\mathrm{bd}(A) | A \in P_w(B) \cup B = P_w(C)\} = \underline{\lim}(C). \end{aligned}$$

The theorem on bounds of positions implies that after a backward step from C we have $\mathrm{Bd}(C) = \mathrm{sc}(C)$. Thus improving moves lead to positions with scores that are determined by the search. We call these *pseudocritical* positions.

If (B, C) is a refutation move, we have after a backward step from C

$$\mathrm{Bd}(C) = \mathrm{Bd}(B) \geq \overline{\lim}(B) = \overline{\lim}(C).$$

Accordingly, after a backward step from C we have, by the theorem on bounds of positions,

$$\mathrm{sc}(B) \geq \overline{\lim}(B) \text{ and } \mathrm{sc}(C) \geq \overline{\lim}(B) = \overline{\lim}(C),$$

i.e. the position B cannot be reached by the opponent, and the (Black) position C cannot be reached by its own side.

If (B, C) is a bad move we have before a backward step from C either $\mathrm{Bd}(C) \leq \mathrm{Bd}(B)$ or $\mathrm{Bd}(C) \leq \underline{\lim}(B)$. Therefore

$$\mathrm{Bd}(C) \leq \max\{\mathrm{Bd}(B), \underline{\lim}(B)\} = \underline{\lim}(C),$$

and by the theorem on bounds of positions, $\mathrm{sc}(C) \leq \underline{\lim}(C)$, i.e. the position C cannot be reached by the opponent.

The sets E, F, and G consisting respectively of pseudocritical positions, positions unreachable by the opponent, and positions unreachable by their own side, may be resolved into subsets of positions all of the same rank: $E = \bigcup E_i$, $F = \bigcup F_i$, $G = \bigcup G_i$.

We use the lower case characters e_i, f_i, and g_i to denote the numbers of elements in the corresponding subsets E_i, F_i, G_i. The base position A_0 belongs to the critical branch. Accordingly,

$$E_0 = \{A_0\}, \qquad F_0 = G_0 = \varnothing, \qquad e_0 = 1, \qquad f_0 = g_0 = 0.$$

Theorem on the Next Moves. *All moves from a position in the set G_i lead to a position in the set F_{i+1}; moves from positions in the set E_i lead to positions in the set $E_{i+1} \cup G_{i+1}$ and so all moves from positions in the set $E_i \cup G_i$ lead to positions in the set $E_{i+1} \cup F_{i+1}$. At least one move leads from any non-terminal position $B \in E_i$ to a position in the set E_{i+1}, and from any non-terminal position $B \in F_i$ exactly one refutation move leads into the set G_{i+1} (see Figure 13).*

PROOF. If $B \in E_i \cup G_i$ is a non-terminal position and (B, C) is the next move backward at B, and $\mathrm{bd}(B)$ is the bound at B after this backward move and $\mathrm{Bd}(B)$ is its last value, then

$$\mathrm{Bd}(C) \leq \mathrm{bd}(B) \leq \mathrm{Bd}(B) < \overline{\lim}(B) = \overline{\lim}(C).$$

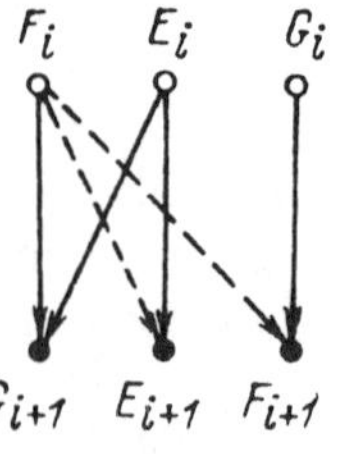

Figure 13

The position C belongs to $E_{i+1} \cup G_{i+1}$ since it is a Black position and $\mathrm{Bd}(C) < \overline{\lim}(C)$. Moreover, on no return to B can the pruning rule be applied, so that all immediate successors of B belong to this set as well. If $B \in G_i$, even stronger conditions are satisfied. After a forward step at B, $\mathrm{bd}(B)$ takes on the value $-\infty < \underline{\lim}$. If before some backward step to B we have $\mathrm{bd}(B) \leq \underline{\lim}(B)$, then

$$\mathrm{Bd}(C) \leq \mathrm{Bd}(B) \leq \underline{\lim}(B) = \max\{\mathrm{bd}(B), \underline{\lim}(B)\} = \underline{\lim}(C),$$

i.e. $C \in F_{i+1}$. If, finally, $B \in F_i$ and (B, C) is the move determining the last value of $\mathrm{Bd}(B)$, then

$$\mathrm{Bd}(C) = \mathrm{Bd}(B) \geq \overline{\lim}(B) = \overline{\lim}(C),$$

i.e. $C \in G_{i+1}$. For any preceding step backward to B from some position C' we have

$$\mathrm{Bd}(C') < \mathrm{Bd}(B) = \overline{\lim}(C').$$

So $C \notin G_{i+1}$ and exactly one move leads from B into the set G_{i+1}. □

Now suppose that the search is optimal, i.e. the leftmost move from any position in the set E is a best move and a move from any position in the set F is a refutation. Then after the first step to the position $B \in E_i$ with the next move (B, C) we have

$$\mathrm{bd}(B) = \max\{-\infty, \mathrm{Bd}(C)\} = \max\{\mathrm{Bd}(C') | (B, C') \in \mathfrak{A}\} = \mathrm{Bd}(B),$$

and after succeeding steps with the next moves (B, C') we have

$$\mathrm{Bd}(C') \leq \mathrm{Bd}(B) \leq \max\{\mathrm{bd}(A) | A \in P_b(C')\} = \underline{\lim}(C),$$

i.e. the moves (B, C) are bad. After the first step backward to a position $B \in F_i$ with next move (B, C) we have $\mathrm{bd}(B) \geq \overline{\lim}(B)$, i.e. the pruning rule must be applied at the next step. Then we must to some extent correct the graph of relationships given in Figure 13 for the sets E_i, F_i, G_i and $E_{i+1}, F_{i+1}, G_{i+1}$. See Figure 14, in which the unique moves available at non-terminal positions are depicted in boldface.

Let us consider three successive ranks, of order $i, i+1, i+2$. Figure 15 shows what moves can be made from positions in the sets E_i, F_i, G_i, E_{i+1}, F_{i+1}, and G_{i+1}. (The dotted lines show the non-considered moves from positions in the set F.) All moves from the set $E_i \cup G_i$ lead into the set $E_{i+1} \cup F_{i+1}$; the moves from the set F_i either lead into G_{i+1} or are ignored.

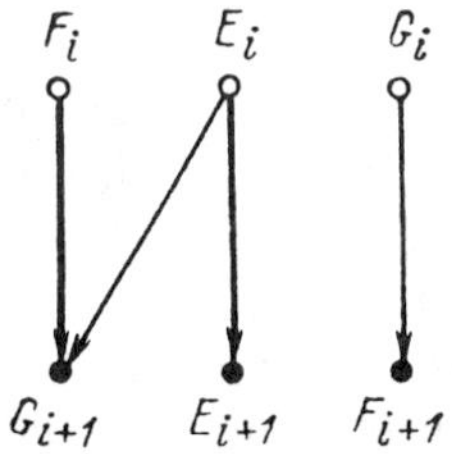

Figure 14

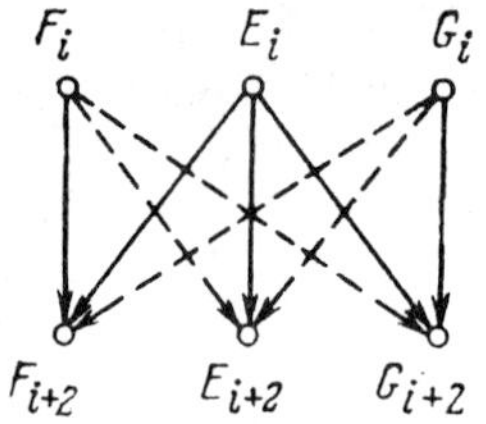

Figure 15

Accordingly the average number of moves from positions in the set $E_i \cup G_i$ is equal to $\mu_i = (e_{i+1} + f_{i+1})/(e_i + f_i)$.

From every non-terminal position in the set $E_{i+1} \cup F_{i+1}$ one move leads into the set $E_{i+2} \cup G_{i+2}$ (from E_{i+1} it leads into E_{i+2}, and from F_{i+1} into G_{i+2}), and positions in the latter set cannot arise in any other way. Therefore the fraction of non-terminal positions in the set $E_{i+1} \cup G_{i+1}$ is equal to

$$\varepsilon_{i+1} = \frac{e_{i+2} + g_{i+2}}{e_{i+1} + f_{i+1}} \leq 1.$$

Thus

$$e_{i+2} + g_{i+2} = \mu_i \varepsilon_{i+1}(e_i + g_i),$$

whence it follows that

$$e_i + g_i = \prod_{j=1}^{[i/2]} \varepsilon_{i-2j+1}\mu_{i-2j},$$

and

$$e_i + f_i = \mu_{i-1} \prod_{i=1}^{[i-1/2]} \varepsilon_{i-2j}\mu_{i-2j-1}.$$

In the optimal search we are now considering, the successor positions are only those in the minimally inclusive W-pruned and B-pruned subtrees $\mathfrak{A}_w$ and $\mathfrak{A}_b$, which intersect in the critical branch W. Only the positions in this branch belong to the set E. Therefore $e_i = 1$ for $i \leq l$, where l is the rank of a terminal position in W, and $e_i = 0$ for $i > l$. If the rules for selecting the next move in the contemplated step are not specifically directed toward the minimization of the μ_i, the values obtained will usually be close to the mean numbers ν_i of all moves from positions in the i-th rank. Moreover, in many interesting cases there are almost no terminal positions of non-maximal rank (we shall see later that this property is shared by many model games played by programs). Then $\varepsilon_i \approx 1$, $\ln \varepsilon_i \approx 0$; we again assume that the values of the $\ln \nu_i$ are approximately equal as between odd and even ranks, and that there is an essential branching in the trees $\mathfrak{A}$, $\mathfrak{A}_w$, and $\mathfrak{A}_b$ for which the number of positions is approximately equal to the total number of terminal positions. Then we can show that the number of positions considered in an optimal search is of the same order of magnitude as the square root of the total number of all positions in the game tree $\mathfrak{A}$. In fact, let h_i be the number of positions of rank i in the tree, let $\overline{\ln \mu}$ and $\overline{\ln \nu}$ denote the means of the $\ln \mu_i$ and $\ln \nu_i$ averaged over all ranks, and let k be the maximum rank in the game tree $\mathfrak{A}$. Then

$$\ln(e_k + f_k) = \ln \mu_{k-1} + \sum_{i=1}^{\lceil (k-1)/2 \rceil} (\ln \varepsilon_{i-2j} + \ln \mu_{i-2j-1})$$

$$\approx \frac{k}{2}\overline{\ln \mu} + O(k) \approx \frac{k}{2}\overline{\ln \nu} + O(k),$$

$$\ln(e_k + g_k) = \sum_{i=1}^{\lceil k/2 \rceil} (\ln \varepsilon_{i-2j+1} + \ln \mu_{i-2j})$$

$$\approx \frac{k}{2}\overline{\ln \mu} + O(k) \approx \frac{k}{2}\overline{\ln \nu} + O(k),$$

$$\ln(e_k + f_k + g_k) \leq \ln(2\max(e_k + f_k, e_k + g_k))$$

$$\approx \frac{k}{2}\overline{\ln \mu} + O(k) \approx \frac{k}{2}\overline{\ln \nu} + O(k),$$

$$\ln h_k = \sum_{i=0}^{k-1} \ln \nu_j = k\overline{\ln \nu},$$

where $\lceil x \rceil$ is the nearest integer not exceeding x (Knuth's notation [19]), and everywhere $O(k) < \lambda k$ for $\lambda \ll 1$. Then

$$h_k = e^{k\overline{\ln \nu}},$$

$$e_k + f_k + g_k \sim e^{\frac{k}{2}\overline{\ln \nu} + O(k)} = \left(\sqrt{n_k}(1+\lambda)\right)^k,$$

where $\lambda \ll 1$.

This conclusion may be sharpened for a completely uniform game $\mathfrak{A}_{m,k}$, in which all positions in the ranks of order $0, 1, \ldots, k-1$ are non-terminal and have the same number m of moves, while the positions of order k are all terminal. For such games

$$e_i = 1,$$

$$f_i = \mu_{i-1} \prod_{j=1}^{\lceil (i-1)/2 \rceil} \varepsilon_{i-2j}\mu_{i-2j-1} - 1 = m^{\lceil (i+1)/2 \rceil} - 1,$$

$$g_i = \prod_{j=1}^{\lceil i/2 \rceil} \varepsilon_{i-2j+1}\mu_{i-2j} - 1 = m^{\lceil i/2 \rceil} - 1, \qquad i = 0, 1, \ldots, k.$$

Thus the number of terminal positions that define the score of the base position, is equal to $\lceil (k+1)/2 \rceil + \lceil k/2 \rceil - 1$, and the number of all positions in the union of the minimal W-pruned and B-pruned subtrees is

$$\frac{m^{\lceil (k+3)/2 \rceil} + 2m^{\lceil (k+2)/2 \rceil} + m^{\lceil (k+1)/2 \rceil}}{m-1} - \left(k + 2 - \frac{2}{m-1}\right).$$

In all, there are $M=(m^{k+1}-1)/(m-1)$ positions in the tree $\mathfrak{A}$. Of course, the minimum number of positions examined in a search is of the order of $\sqrt{M}$ for k even, and $\sqrt{mM}$ for k odd.

If the best and refutation moves do not immediately come to mind, a larger number of positions must be inspected. It can be shown that there are $(c\overline{m})^{\overline{k}/2}$ of these, where $\overline{k}$ is the mean depth of the tree, $\overline{m}$ is the mean number of moves at a generic position and $c>1$ is of the order of the sum of the mean numbers of moves at positions in the set E and the non-refutation moves at positions in the set F. We content ourselves with a definition of the mathematical expectation of the number of positions considered during a search of a completely uniform game $\mathfrak{A}_{m,k}$ when the numbers of improving moves at all positions $B\in E$ are independent random variables with mean γ (clearly $\gamma\geq 1$), and the numbers of improving moves and bad moves at positions $B\in F$ are also independent random variables with means δ and ε. The independence of these random variables implies that the means of the numbers of positions in the trees are summed over the means of the numbers of positions in the subtrees in precisely the same way as individual values of these numbers in any concrete instance.

Let Ω_i be the mathematical expectation of the number of terminal positions in the B-subtree of the search tree, for positions $B\in E_i$ and let Φ_i and Ψ_i be the corresponding quantities for the sets F_i and G_i. Then

$$\begin{aligned}
&\Omega_0=\Phi_0=\Psi_0=1,\\
&\Omega_{i-1}=\gamma\Omega_i+(m-\gamma)\Phi_i,\\
&\Phi_{i-1}=\delta\Omega_i+\varepsilon\Phi_i+\Psi_i,\\
&\Psi_{i-1}=m\Phi_i,\qquad i=1,2,\ldots,k.
\end{aligned}$$

The system of $3k$ equations represented by the last three lines above is linear and homogeneous in $3k+3$ unknowns, and has rank $3k$. Accordingly it suffices to find three linearly independent solutions of the system. These we seek in the form

$$\Omega_i=\Omega t^{k-i},\qquad \Phi_i=\Phi t^{k-i},\qquad \Psi_i=\Psi t^{k-i}.$$

Substituting these values yields the three equations

$$\begin{aligned}
&\Omega t=\gamma\Omega+(m-\gamma)\Phi,\\
&\Phi t=\delta\Omega+\varepsilon\Phi+\Psi,\\
&\Psi t=m\Phi.
\end{aligned}$$

For these equations to have a non-zero solution the determinant

$$\begin{vmatrix} t-\gamma & -(m-\gamma) & 0\\ -\delta & t-\varepsilon & -1\\ 0 & -m & t\end{vmatrix}
=t^3-(\gamma+\varepsilon)t^2-[m(1+\delta)-\gamma(\delta+\varepsilon)]t+\gamma m$$

must vanish.

Since $1 \le \gamma \le m$ one of the roots, say t_1, of the equation

$$t^3 - (\gamma + \varepsilon)t^2 - [m(1+\delta) - \gamma(\delta + \varepsilon)]t + \gamma m = 0$$

must lie in the half-open interval $(0, \gamma]$. We write

$$q\delta = \frac{m - \gamma}{m + \varepsilon t_1 - t_1^2}\delta = \frac{\gamma - t_1}{t_1}.$$

In the cases that interest us, γ, δ, and ε are much smaller than m. Then t_1 and q depend on δ as shown in Figure 16, and $q \approx 1$. The value of t_1 is conveniently expressed in terms of q:

$$t_1 = \frac{\gamma}{1 + q\delta}.$$

The second and third roots are solutions of the quadratic equation

$$t^2 - (\gamma + \varepsilon - t_1)t - \frac{\gamma m}{t_1} = 0$$

and are equal to

$$t_{2,3} = \frac{\gamma q\delta + \varepsilon(1 + q\delta)}{2(1 + q\delta)} \pm \sqrt{m(1 + q\delta) + \left(\frac{\gamma q\delta + \varepsilon(1 + q\delta)}{2(1 + q\delta)}\right)^2}.$$

Thus the general solution of our system of equations has the form

$$\Omega_i = \Omega^{(1)} t_1^{k-i} + \Omega^{(2)} t_2^{k-i} + \Omega^{(3)} t_3^{k-i},$$
$$\Phi_i = \Phi^{(1)} t_1^{k-i} + \Phi^{(2)} t_2^{k-i} + \Phi^{(3)} t_3^{k-i},$$
$$\Psi_i = \Psi^{(1)} t_1^{k-i} + \Psi^{(2)} t_2^{k-i} + \Psi^{(3)} t_3^{k-i},$$

where

$$\Phi^{(1,2,3)} = \frac{t_{1,2,3} - \gamma}{m - \gamma}\Omega^{(1,2,3)},$$

$$\Psi^{(1,2,3)} = \frac{m}{m - \gamma}\,\frac{t_{1,2,3} - \gamma}{t_{1,2,3}}\Omega^{(1,2,3)}.$$

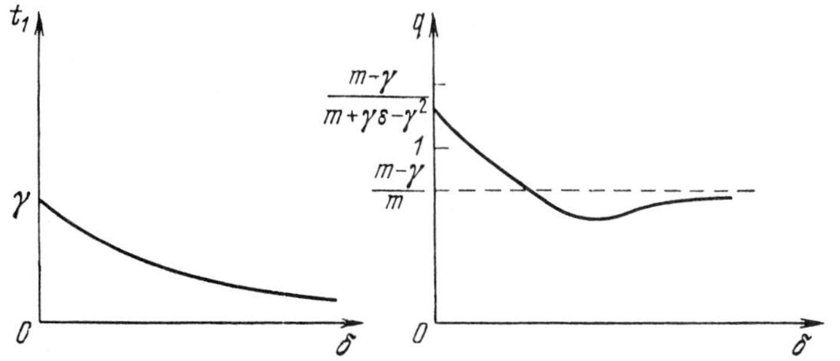

Figure 16

The values $\Omega^{(1,2,3)}$ may be found from the equations $\Omega_k = \Phi_k = \Psi_k = 1$.

Thus the mathematical expectation of the number of terminal positions considered in the tree $\mathfrak{A}_{m,k}$ is equal to

$$\Omega_0 = \frac{(m-t_2)(m-t_3)}{m(t_1-t_2)(t_1-t_3)} t_1^{k+1} + \frac{(m-t_1)(m-t_3)}{m(t_2-t_1)(t_2-t_3)} t_2^{k+1} + \frac{(m-t_1)(m-t_2)}{m(t_3-t_1)(t_3-t_2)} t_3^{k+1}.$$

CHAPTER 2

Heuristic Methods

Control of the Tree Size and Evaluation Functions

The goal of a game-playing program is to recommend a move in every position presented to it. The recommended move need not be the best, for instance in the sense defined in the preceding chapter, but it should have the highest quality possible, as seen by the professional player, and must be chosen within a preselected time limit. An arbitrary algorithm for choosing a move in the positions of a given game may be looked on as an algorithm for choosing the best move in the base position of the game tree of another game, which we shall call a *model* game. The tree for the model game is a subtree of the original game tree.

In constructing the model game we may adopt the goal of finding, so far as we can, the best move for the original game, or we may be guided by other considerations mentioned earlier. In practically all research on game-playing programs it has been assumed that one is to find the best move. Either an exhaustive search or a pruned search of the game tree may be used. In the course of such an algorithm, or its analogues, one constructs W-pruned and B-pruned subtrees of the contemplated game tree with their roots at the base position. Thus their processing time is of an order of magnitude not less than the number of positions in these subtrees. We are interested in games with strongly branching trees, such that even with maximal pruning the search cannot be completed within the available time. There are two ways to save time: a) build as small a model game as possible (with as few positions as possible in the subtrees to be studied), or b) perfect the method of choosing the best move for a fixed game tree.

We shall discuss methods based on the use of meaningful properties of the contemplated games. These methods allow us to choose moves that are more or less satisfactory. If an algorithm for an arbitrary game is to be significantly more effective than a pruned search, it must not construct W-pruned and B-pruned subtrees, except perhaps in a preliminary and implicit way. For such an algorithm we must define precisely what we mean by calling a chosen move *fairly good*. If the algorithm is deterministic, then by using the theorem on extensions of game trees we can find a game such that in its winning positions one will always win the game against errorless play by the opponent (if the game includes random factors, substitute almost always for always). Of course, there exist concrete games to which the notion of a *fairly good move* cannot be applied.

We shall call a method *exact* when it can be shown to solve the assigned task correctly, and *heuristic* when no such proof has yet been found. (The term *heuristic* is used in various senses; we have chosen one that suits us.) The exact method constructs a model equivalent to the original game, i.e. the best moves in the base positions of its trees are best moves in the corresponding original game. A method for accelerating the choice of a best move in the base position of a fixed game tree may be called exact if a proof exists that its application does in fact shorten the search, i.e. decrease the number of positions examined.

We begin with the heuristic methods that up to now have been the most thoroughly elaborated. The first of these was described by Shannon [39]. In a game tree with its root at a given position, he proposed to consider only positions in the low ranks; that is, Shannon's algorithm has a parameter n, the depth of the search, which is the maximum rank of the positions to be examined. All positions in the n-th rank are considered terminal. We must specify their scores; these of course need not coincide with the true scores, which we do not know. We select an ensemble of features to be evaluated, from among those occurring in the various theories of the given game. Shannon's paper discusses a chess program and such features as the material balance, weak and strong squares, attacks by the pieces on them, elements of the pawn structure, etc.

The score for a terminal position in the model game must be equal to the value of some function, that is reasonably easy to compute, of the features characterizing the position. (Of course, an actual terminal position in the original game is terminal in the model also; the won and lost positions have model scores denoting won and lost games; a drawn position has an intermediate score.) The contemplated function can be computed at any position in the game; we call it an *evaluation function*. Let us now compare models having varying depths of search but the same scoring function. Since the game tree in chess is finite, Shannon's game tree with sufficiently large n will coincide with it. In that case the model will choose the actual best move.

On the other hand, if we could define an evaluation function that would yield the true score in every position, the model need have only the depth $n = 1$. This is often referred to as looking ahead one ply (taking account of the fact that a whole move consists of the union of a move by White and a move by Black). What significance can we attach to a search with depth $n > 1$ that does not reach the end of the game? Wins and losses that occur in very few moves, the so-called catastrophes, will of course be examined. But this is wholly insufficient for what would be called decent play. If the evaluation function can only distinguish between a position after a catastrophe has occurred and a prior position, it is of no use for anything else. However, for certain postulates on the character of the game and on the properties of the evaluation function one can prove that even in the absence of catastrophes the quality of play increases monotonely with the depth of the search.

Using Shannon's model, we now describe a somewhat idealized game and develop a naive probabilistic basis for an effective choice of best move in its positions. Let $\mathfrak{A}$ be a completely uniform game, i.e. White and Black move alternately, at every position the number m of available moves is the same, and all terminal positions have the same (sufficiently large) rank N. Suppose, moreover, that the number of winning moves at every position where they exist is the same, and is equal to 3.Let the terminal positions have scores 0 or 1. We postulate that the evaluation function $f(A)$ is a random variable such that in a White (Black) position won for White (Black) we have $f(A)=1$ (0) with probability p, and $f(A)=0$ (1) with probability $(1-p)$; in lost positions the corresponding probabilities are $q < p$ and $(1-q)$. At different positions A the values of $f(A)$ are independent random variables. It is easy to see that in this game there are no catastrophes.

As usual, we assume that White is to move in the base position A_0. A game-playing program using Shannon's model will make an errorless move whenever the position B arising in the game $\mathfrak{A}$ after a winning move has the score 1 and all positions B' to which losing moves (A_0, B') lead have score 0. We must determine the probabilities of the scores 1 and 0 for winning and losing moves of rank 1 in the original game $\mathfrak{A}$, using the Shannon model of depth n.

Let us begin with a Shannon model of depth 1 (the search omits some of the positions B of rank 1 whose scores are compared in searching for the best move). If for a position B of rank 1

$$\operatorname{msc}(B) = \min\{f(C) \mid (B,C) \subset \mathfrak{A}\} = 1,$$

the values of the evaluation function $f(C)$ are equal to 1 for all immediate successors C of B. Because these random variables are independently distributed for different positions C (as shown in Figure 17 the $C_1, C_2, \ldots, C_m$ are immediate successors of B), the probability of this event is equal to the product of the probabilities for every event $f(C_i) = 1$:

$$\mathbf{P}(\min\{f(C) \mid (B,C) \in \mathfrak{A}\} = 1)$$
$$= \mathbf{P}(f(C_1 = 1)) \cdot \mathbf{P}(f(C_2) = 1) \cdots \mathbf{P}(f(C_m) = 1).$$

We note that all the $C_1, C_2, \ldots, C_m$ are Black positions.

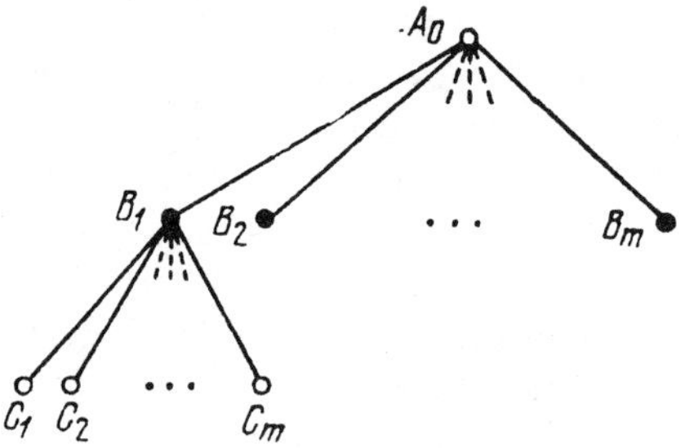

Figure 17

When B is a won position for White in the original game we shall denote the corresponding probability by $1-Q_1$, otherwise by $1-P_1$ (it is natural to assume that $P_0=p$, $Q_0=q$). In the first case all the positions $C_1, C_2, \dots, C_m$ are also won for White. Accordingly,

$$1-Q_1=p^m.$$

In the second case, with s positions won for Black and the rest for White, we have

$$1-P_1=p^{m-s}q^s.$$

Let P_n be the probability of obtaining a score 1 for the base position A_0 in Shannon's model game when A_0 (a White position, as always) is won for White in the original game, and let Q_n be the probability of obtaining the score 1 if it is lost for White. Then the probability that the positions C_i $(1 \le i \le m)$ in Figure 17 will obtain the score 1 in a Shannon model with depth $n+1$ is P_n if C_i is in fact a won position and Q_n if it is a lost position. As in the model of depth 2,

$$1-Q_{n+1}=P_n^m,$$
$$1-P_{m+1}=P_n^{m-s}Q_n^s.$$

If P_n is near enough to 1 and Q_n to 0, P_{n+1} will be even larger and Q_{n+1} even smaller. Thus the probability of obtaining a correct score for the positions, and so finding the best move, increases with the depth n of the search and tends to 1 so that it is practically equal to 1 for $n \ll N$. Table 1 displays the values of P_n and Q_n as functions of n for $m=10$, $s=2$, $p=0.99$, $q=0.1$. However, if p is only slightly smaller than 0.99 or q is somewhat larger than 0.1, both P_n and Q_n fail to converge. After some number of oscillations both tend to 1 for one parity of n and to 0 for the other parity.

This result—the increase in the reliability of the Shannon model decision as the depth of the search increases—follows from an important property of the games in question, namely that in a won position there are more than several winning moves and that we can construct a relatively simple evaluation function that correlates well with the true score. Seemingly, both these properties are necessary if we are to count on a good result from

Table 1. Probability of Correct Model Score for Shannon Model Games

Notation

m—number of moves at non-terminal positions
s—number of winning moves at winning non-terminal positions
n—depth of the model
P_n—probability of a win in the model game for a won position in the original game
Q_n—probability of a win in the model game for a lost position in the original game

$m = 10, \quad s = 2$

n	P_n	Q_n	P_n	Q_n	P_n	Q_n
0	0.9900	0.1000	0.9900	0.1200	0.9880	0.1000
1	0.9908	0.0956	0.9867	0.0956	0.9909	0.1137
2	0.9915	0.0885	0.9918	0.1252	0.9880	0.0872
3	0.9927	0.0817	0.9853	0.0792	0.9931	0.1139
4	0.9937	0.0709	0.9944	0.1374	0.9877	0.0669
5	0.9952	0.0612	0.9819	0.0543	0.9959	0.1162
6	0.9964	0.0467	0.9974	0.1666	0.9869	0.0398
7	0.9979	0.0355	0.9728	0.0252	0.9986	0.1234
8	0.9988	0.0210	0.9995	0.2410	0.9850	0.0142
9	0.9996	0.0123	0.9421	0.0051	0.9998	0.1407
10	0.9998	0.0044	0.9998	0.4490	0.9802	0.0018

applying Shannon's model, or others like it that we shall describe in Section 3 below. It may turn out that for some games the fulfillment of these conditions would be strictly proved or disproved. As of now such proof is lacking and we must resort to statistical tests of their fulfillment.

We should not be discouraged by the stringent requirements placed on the parameters p and q that define the correlation between the scores and the evaluation function. These requirements stem from the simplifications we have adopted, in particular from the postulate that the evaluation functions evaluated before and after a move are independently distributed, and from the non-differentiability of their values (0 and 1 only). Moreover, we must keep in mind that one purely qualitative achievement in the construction of the evaluation function, namely its positive correlation with the true score, may fail to yield any advantage (also, the required depth of the search depends on the level of this correlation and on circumstances that are not random). We shall return to this question in Chapter 4.

The evaluation function may depend on factors other than the true scores of the corresponding positions. If at some position a win can be achieved in 10 moves, and at another position in 4 moves, the scores of the two positions are identical. Yet for the players the two positions are not equivalent; it is always best to choose a move leading to a desired result in the shortest possible time. Therefore we may define the evaluation function in such a way that conditions near the end of the game can increase the score (when we expect a win). It is especially important to have such values for the evaluation function at positions where the theory of the game specifies a win (this does not mean that we can give a strict proof of a win).

For instance, in easily won chess endgames, where a lone King is to be mated, one may avoid a failure to win, so long as the 50 move rule and the rule of triple repetition of a position do not apply. However, even when the evaluation function agrees with the true score, it is impossible to guarantee a mate when the search depth $n \leq 10$. At the same time, if the evaluation function yields a better value for the attacking side when the lone King is near the edge of the board and the corners (when the mate is to be given by Bishop and Knight the corner must be of the same color as the Bishop), and the attacking King is near the opposing King, then a program with depth 4 will win, and the fourth ply is needed only to test whether a mate or a stalemate occurs on the third ply. For a mate by two minor pieces a specially selected score is needed for a few standard positions with the opposing King in or near a corner of the board.

As we have said, an evaluation function should be relatively easy to compute. The simplest would be linear. We shall define a collection of features of a position—predicates $p_i(A)$ which have the value 1 if A possesses the i-th feature and the value 0 otherwise. If q_i is the weight of the i-th feature ($i = 1, 2, \ldots, N$) the evaluation function is defined by the formula

$$\Phi(A) = \sum_{i=1}^{N} q_i p_i(A),$$

where the weights may be either positive or negative. It can be shown that when the number of features to be taken into account is large, then even for a very simple game there exists no linear evaluation function that is monotonely related to the true scores of the positions [22]. Therefore the problem of complicated methods for computing the evaluation function is unavoidable. However, the determination of model scores of positions by Zermelo's formula, starting from the value of an evaluation function in pseudoterminal positions, is also a method for computing evaluation functions and we may confine our considerations to processes that require significantly less work.

The use of threshold logic provides an example of a moderate complication of the methods for computing an evaluation function. Aside from the initial set of features—the system of predicates $\{p_i(A)\}_1^n$—we consider a set of evaluation functions

$$\left\{ \Phi_j(A) := \sum_{i=1}^{N} q_i^j p_i(A) \right\}_1^M$$

and thresholds R_j for them. The evaluation function $\Phi(A)$ is defined by the formula

$$\Phi(A) = \sum_{j=1}^{M} Q_j \Pi_j(A),$$

where $\{Q_j\}_1^M$ is a set of weights and

$$\Pi_j(A) := \begin{cases} 1, & \text{if } \Phi_j(A) = \sum_{i=1}^{M} q_i^j p_i(A) \geq R_j, \\ 0, & \text{if } \Phi_j(A) < R_j, \qquad j = 1, 2, \ldots, M. \end{cases}$$

The ability of threshold logic to represent essential properties of certain objects of study was investigated by Minsky and Papert [25], who elucidated the constraints on the nature of the properties, the structure of the logical scheme, and the number of elements. However, the properties they studied that are to be reflected by threshold logic are quite different from those properties of a position that are to be represented by an evaluation function.

In the existing game programs the approach to the use of threshold logic to compute the evaluation functions $\Phi(A)$ is somewhat different. It is connected with the concept of the *position type* borrowed from chess theory. An elementary feature of a type is computed by means of the value of a linear function

$$\Psi_j(A) := \sum_{i=1}^{N} \psi J_i^j p_i(A)$$

of the elementary features $p_i(A)$ of a position A and the thresholds R_i $(i = 1, 2, \ldots, M-1)$. The position type is defined by logical functions of these features. For example, the predicates $P_k(A)$ indicating that A is of type k may be defined by the formulae

$$P_1(A) := \{\Psi_1(A) \geq R_1\},$$

$$P_k(A) := \mathop{\&}_{j=1}^{k-1} \{\Psi_j(A) < R_j\} \& \{\Psi_k(A) \geq R_k\},$$

$$k = 2, \ldots, M-1,$$

$$P_M(A) := \mathop{\&}_{j=1}^{M-1} \{\Psi_j(A) < R_j\} \& \{\Psi_M(A) \geq R\}.$$

For every type a linear evaluation function

$$\Phi_j(A) := \sum_{i=1}^{N} \phi_i^j P_i(A), \; j = 1, 2, \ldots, M,$$

is defined; the functions ϕ_i^j of the features $p_i(A)$ with non-zero coefficients in this evaluation function are usually not the same as those with non-zero coefficients in the functions Ψ_j that define the type. Thus the value of the evaluation function for an arbitrary position A is given by

$$\Phi(A) = \sum_{j=1}^{M} \Phi_j(A) P_j(A).$$

Another way to develop a complex evaluation function, still within the framework of linear functions, is to use not only elementary features given *a priori* but also logical functions of them. An arbitrary function of the elementary predicates $P_i(A)$ $(i=1,2,\ldots,N)$ defined above can be defined in the same way. In fact, let $\Phi_{\varepsilon_1\varepsilon_2\ldots\varepsilon_N}$ be the value of the contemplated function for the arguments $p_i(A)=\varepsilon_i$, where ε takes on the values 0 or 1; then

$$\Phi(A) = \sum_{\varepsilon_1=0}^{1} \sum_{\varepsilon_2=0}^{1} \cdots \sum_{\varepsilon_N=0}^{1} \Phi_{\varepsilon_1\varepsilon_2\cdots\varepsilon_N}\{p_1(A) = \varepsilon_1\} \& \cdots \& \{p_N(A) = \varepsilon_N\}.$$

In such a representation the number of terms is very large (equal to 2^N), and only relatively short formulae may be easily computed. In practical applications the evaluation functions usually do have non-elementary terms, but only a few of them.

So, in actual game programs the evaluation functions either depend linearly on features of a position that are sufficiently simple in their definition or they are superpositions of linear and logical functions. In constructing a model game we must decide what features to take into account, and what weights to assign to them. Shannon proposed deriving a collection of elementary features from the theory of any specific game. In the theories of chess, checkers (draughts), bridge, etc. there are a series of such features that monotonely influence the outcome of the game. Almost all of them, however, lack a precise formal definition (material balance is an exception). We need to construct formal features that coincide at least approximately with those we are interested in. It is important to note that essential features have been found which are unknown in the theories of actual games. Also, there are interesting features for which no algorithmically expressible equivalents (even approximate) have been found.

Many papers have been devoted to the problem of weights to be assigned to features, and two basic concepts have been elaborated. The first says that in every position there is one dominant feature, and the weights should be so chosen that the presence of an arbitrary positive feature would be suppressed only by the presence of a more important feature, and not by a combination of less important features. If an arbitrary logical function of the elementary features may be thought of as a new feature that affects the value of the evaluation function, the first concept is meaningless. Moreover, when the amount of computing is limited and therefore there are not too many terms in the linear expression, the concept in its pure form conflicts with the empirically based method of computing the material balance (the values of the pieces in chess or bridge). Attempts to apply it even to positional features have not succeeded (the strong game of Samuel's program [33] for American checkers would appear to be unconnected with the fact that he used this concept in assigning weights to features).

The second concept approaches the assignment of weights empirically. They are introduced as parameters in the program and after various games have been played the values chosen are those for which the program plays most strongly. It has also been proposed that an automatic process be set up, in which the values of the parameters change in the course of a single game. Such programs are sometimes referred to as *self-learning*, but the appropriateness of this term is disputable. The authors have made a statistical study of the way in which various values of the weights affect the strength of a chess program. It turns out that random variation of the weights, within wide limits around the values established *a priori* on the basis of certain chess-theoretical considerations, has very little influence on the strength of the play (of course, we are speaking of positional features). When, however, the weights are so chosen that some positional features have practically no influence on the score, the quality of play worsens significantly. It is also important to note that no positional factor will dominate a group containing a substantial number of other features.

Some studies, e.g. [12], have taken up the problem of automatically choosing the logical functions (of the elementary features) from which the evaluation function is built up. The procedure is to enumerate the logical functions, compute the correlation of these with scores made by experts for various test positions, and choose the most informative of the logical functions. The work done by such a program may be more rightfully called learning than can changing the weights of features in a given program. The *teacher*—the expert who specifies the collection of test positions and their scores—plays a fundamental role in this process. Such methods have achieved definite results in studies on pattern recognition (cf. [6]). However, the learning of programs for the diagnosis of position scores in a game has not yet yielded any substantial results.

The first functioning chess programs implemented Shannon's ideas literally and played very weakly. The search was comparatively shallow, many technical programming problems were solved in a far from satisfactory fashion, and the pruning method (the α, β heuristic) was not used. A much more salient shortcoming of the programs was their use of a fixed search depth and their failure to vary it depending on the character of the position. For instance, an exchange might be broken off at an arbitrary instant when the material balance on the board failed to reflect the true balance. The shortcomings of the first models were removed in various ways, some of which will be discussed in the next section. No really strong chess program has yet been developed, although very strong programs exist for simpler games. A program with rather simple ideas plays Russian draughts [36–38] at a good level, and Samuel's program for American checkers (mentioned earlier) plays at a level that need not yield to professionals. There also exist programs that correctly solve the problem of mate in a few moves or play correctly in simple endgames such as King and Pawns against King.

On the Order in which Positions are Searched in the Game Tree

Let us begin with algorithms for a search with pruning (the α, β heuristic). As we proved in Chapter 1, the minimum number of positions to be examined in the game tree $\mathfrak{A}$ in order to establish the score of the base position A_0 and the best initial move, is of the order of $m^{k/2}$, where m is the mean number of moves at the positions to be examined and k is the depth of $\mathfrak{A}$. To attain this minimum we must examine the best move first when in the critical path W, and we must examine one of the refutation moves first in positions that are unreachable by the opponent. If, on the other hand, we examine the moves at any given position in increasing order of quality, i.e. each move examined is better than its predecessor, we will be making a practically exhaustive search and will look at $O(m^k)$ positions. It is therefore essential to determine a sequence for looking at positions that is as close as possible to the optimal.

However, in order to guarantee optimality of the search sequence, we need to know which is the best move at most positions in the game tree $\mathfrak{A}$, i.e. we must be able to solve at these positions the same problem we needed to solve only at the base position A_0. We are interested in the case in which we do not know how to solve this problem and can only more or less rapidly guess the necessary best or refutation move. The search process itself may be looked on as a test of the correctness of our guesses. The methods for guessing may be formulated as a sharpening of the rule for deepening a search with pruning, which tells us how to choose the next move in a step forward.

Since the search sequence to be constructed cannot be guaranteed to be optimal, there may be errors in it. Therefore it is useful to know how the number of positions in the search depends on errors of different kinds and what errors we should above all try to avoid, even at the cost of an increase in the number of errors of other kinds. In Section 3 of Chapter 1 the quality of a search sequence was described by three parameters:

γ—the mean number of improving moves at positions with determined scores (candidates for the critical branch);

δ—the mean number of improving moves at positions from which later refutation moves can be found (unreachable by the opponent);

and

ε—the mean number of bad moves at the same positions.

In a completely uniform game $\mathfrak{A}_{m,n}$ of depth n with m moves at each non-terminal position, when the numbers of improving and bad moves are equal to the mean values corresponding to the position type, the number of terminal positions in the search tree is defined by the formula

$$\Omega_{m,n} = \Omega^{(1)} t_1^n + \Omega^{(2)} t_2^n + \Omega^{(3)} t_3^n,$$

where $t_1, t_2, t_3, \Omega^{(1)}, \Omega^{(2)}, \Omega^{(3)}$ are determined by the parameters $\gamma, \delta, \varepsilon$, and m. Thus the way in which the growth of the number of positions depends on the depth n of the tree is determined by the largest of the numbers t_1, t_2, t_3; these are the roots of the cubic equation

$$t^3 - (\gamma + \varepsilon)t^2 - (m(1+\delta) - \gamma(\delta + \varepsilon))t + \gamma m = 0.$$

If $\gamma, \delta, \varepsilon$ are much less than m, the roots t_1, t_2, t_3 can be expressed in terms of a parameter q which is near to 1:

$$t_1 = \frac{\gamma}{1 + q\delta},$$

$$t_{2,3} = \pm \sqrt{m(1+q\delta) + \frac{\left(\frac{q\gamma\delta}{1+q\delta} + \varepsilon\right)^2}{4}} + \frac{\frac{q\gamma\delta}{1+q\delta} + \varepsilon}{2}.$$

$$0 < t_1 \le \gamma,$$

$$|t_2| = \sqrt{m(1+q\delta)} + O(\gamma + \varepsilon),$$

$$|t_3| = \sqrt{m(1+q\delta)} - O(\gamma + \varepsilon).$$

The number of positions to be examined increases with increasing values of $\gamma, \delta, \varepsilon$ but these parameters have differing effects on the rate of increase. Roughly speaking, γ and ε form additive terms, whereas δ determines the multiplier $\sqrt{1+q\delta}$ in the formula for t_2, which specifies the order of the growth rate in the number of terminal positions as it depends on the depth n of the tree $\mathfrak{A}_{m,n}$. The values of $\Omega_{m,n}$ are shown in Table 2 for various values of γ, δ, and ε.

Now let us ask how individual errors in determining the search sequence in the search tree $\tilde{\mathfrak{A}}$ influence the number of positions. Let A be a next position of rank $l < n$ which is unreachable by the opponent. If instead of examining the refutation move (A, B) we choose the improving move (A, B'), we include the B'-subtree in the tree $\tilde{\mathfrak{A}}$ to define the score at B'. This is equivalent to the search tree of the game $\mathfrak{A}_{m,n-l-1}$ and, as we found in §3 of Chapter 1, in the absence of other errors this tree contains

$$\frac{\sqrt{m}^{\,n-l-2}}{2}\left((\sqrt{m}+1)^2 + (-1)^{n-l}(\sqrt{m}-1)^2\right) - 1$$
$$= m^{\lceil (n-l)/2 \rceil} + m^{\lceil (n-l-1)/2 \rceil} - 1$$

Table 2. Number of Positions in the Search Tree as a Function of the Number of Moves at Non-Terminal Positions and the Move-Inspection Sequence

Notation

m—number of moves at non-terminal positions,
n—depth of the search tree,
γ—mean number of improving moves at candidates for the critical branch,
δ—mean number of improving moves at positions unreachable by the opponent,
ε—mean number of bad moves at positions unreachable by the opponent,
A_n—number of terminal positions in the tree $\mathfrak{A}_n$ that defines the score of the base position,
B_n—number of terminal positions in the tree $\mathfrak{B}_n$, which determines that the base position is unreachable by the opponent,
C_n—number of terminal positions in the tree $\mathfrak{C}_n$, which determines that the base position is unreachable by oneself.

$\gamma = 1, \quad \delta = 0, \quad \varepsilon = 0$

	$m = 2$			$m = 10$			$m = 40$		
n	A_n	B_n	C_n	A_n	B_n	C_n	A_n	B_n	C_n
0	1	1	1	1	1	1	1	1	1
1	2	1	2	10	1	10	40	1	40
2	3	2	2	19	10	10	79	40	40
3	5	2	4	109	10	100	1639	40	1600
4	7	4	4	199	100	100	3199	1600	1600
5	11	4	8	1099	100	1000	65599	1600	64000
6	15	8	8	1999	1000	1000	127999	64000	64000
7	23	8	16	10999	1000	10000	2623999	64000	2560000

$\gamma = 2, \quad \delta = 0, \quad \varepsilon = 0$

	$m = 2$			$m = 10$			$m = 40$		
n	A_n	B_n	C_n	A_n	B_n	C_n	A_n	B_n	C_n
0	1	1	1	1	1	1	1	1	1
1	2	1	2	10	1	10	40	1	40
2	4	2	2	28	10	10	118	40	40
3	8	2	4	136	10	100	1756	40	1600
4	16	4	4	352	100	100	5032	1600	1600
5	32	4	8	1504	100	1000	70864	1600	64000
6	64	8	8	3808	1000	1000	202528	64000	64000
7	128	8	16	15616	1000	10000	2837056	64000	2560000

Table 2 (*Continued*)

$\gamma = 1, \quad \delta = 1, \quad \varepsilon = 0$

	$m = 2$			$m = 10$			$m = 40$		
n	A_n	B_n	C_n	A_n	B_n	C_n	A_n	B_n	C_n
0	1	1	1	1	1	1	1	1	1
1	2	2	2	10	2	10	40	2	40
2	4	4	4	28	20	20	118	80	80
3	8	8	8	208	48	200	3238	198	3200
4	16	16	16	640	408	480	10960	6438	7920
5	32	32	32	4312	1120	4080	262042	18880	257520
6	64	64	64	14392	8392	11200	998362	519562	755200
7	128	128	128	89920	25592	83920	21261264	1753562	20782480

$\gamma = 1, \quad \delta = 0, \quad \varepsilon = 1$

	$m = 2$			$m = 10$			$m = 40$		
n	A_n	B_n	C_n	A_n	B_n	C_n	A_n	B_n	C_n
0	1	1	1	1	1	1	1	1	1
1	2	2	2	10	2	10	40	2	40
2	4	4	4	28	12	20	118	42	80
3	8	8	8	136	32	120	1756	122	1680
4	16	16	16	424	152	320	6514	1802	4880
5	32	32	32	1792	472	1520	76792	6682	72080
6	64	64	64	6040	1992	4720	337390	78762	267280
7	128	128	128	23968	6712	19920	3409108	346042	3150480

$m = 10$

$\gamma = 1.5, \quad \delta = 0.5$

	$\varepsilon = 0.5$			$\varepsilon = 1.0$		
n	A_n	B_n	C_n	A_n	B_n	C_n
0	1	1	1	1	1	1
1	10	2	1	10	2	10
2	32	16	20	36	18	25
3	184	44	160	203	61	175
4	650	274	440	820	337	606
5	3304	902	2740	4096	1353	3372
6	12623	4843	9020	17648	6773	13534
7	60100	17753	48430	84046	29132	67734

Table 2 (*Continued*)

$\gamma = 1.5, \quad \delta = 1$

	$\varepsilon = 0.5$			$\varepsilon = 1.0$		
n	A_n	B_n	C_n	A_n	B_n	C_n
0	1	1	1	1	1	1
1	10	2	10	10	3	10
2	36	21	25	40	23	30
3	235	72	212	256	94	230
4	963	483	719	1179	580	935
5	5554	1924	4834	6697	2694	5798
6	24685	11351	19239	32943	15188	26939
7	133508	49599	113507	178512	75069	151879

$\gamma = 2, \quad \delta = 0.5$

	$\varepsilon = 0.5$			$\varepsilon = 1.0$		
n	A_n	B_n	C_n	A_n	B_n	C_n
0	1	1	1	1	1	1
1	10	2	10	10	2	10
2	36	16	20	40	18	25
3	200	46	160	220	62	175
4	768	283	460	940	348	625
5	3800	986	2830	4660	1442	3475
6	15484	5223	9855	20860	7248	14425
7	72750	20208	52228	99700	32102	72475

$\gamma = 2, \quad \delta = 1$

	$\varepsilon = 0.5$			$\varepsilon = 1.0$		
n	A_n	B_n	C_n	A_n	B_n	C_n
0	1	1	1	1	1	1
1	10	2	10	10	3	10
2	40	21	25	44	23	30
3	250	76	212	272	97	230
4	1105	500	756	1320	599	970
5	6212	2111	5003	7432	2889	5990
6	29316	12271	21114	37976	16311	28890
7	156803	56566	122713	206440	83177	163110

terminal positions. If we first examine the bad move (A, B'') instead of the move (A, B), then in the absence of other errors we must also examine the minimal W- and B-subtrees that bound the score of the position B'' from one side. These contain

$$\frac{\sqrt{m}^{\,n-l-2}}{2}\left((\sqrt{m}+1)+(-1)^{n-l}(\sqrt{m}-1)\right) = m^{\lceil (n-l-1)/2 \rceil}$$

terminal positions, i.e. a lesser number.

Now suppose that as a result of the search the next position A turns out to be a candidate for the critical branch. After each improving move (A, B) we must determine the score of the position B, i.e. in the absence of errors in the search of the B-subtree of the game $\mathfrak{A}_{m,n}$ we must consider

$$m^{\lceil (n-1)/2 \rceil} + m^{\lceil (n-l-1)/2 \rceil} - 1$$

terminal positions; after a bad move (A, B) we must consider, in the best case, $m^{\lceil (n-l-1)/2 \rceil}$ such positions (l is the rank of the position A). If a bad move (A, B') is examined by mistake before an improving move, it may appear to be improving. In that case, in the absence of other errors, we must consider $m^{\lceil (n-l)/2 \rceil}$ superfluous terminal positions, i.e. fewer than if we had preferred an improving move to a refutation move for examination among the positions of the same rank. Moreover, when the number of errors in the search sequence is not very large, the number of candidates for the critical branch is much less than the number of positions unreachable by the opponent, and when the probabilities of error in positions of the first type are identical, the number will still be less.

Thus, the lower the rank of the position in which an error in defining the search sequence occurs, the more serious the consequences of the error. Therefore some extra work is worth doing in sharpening the search sequence in next positions of low rank. On the average, first place should be given to attempts to find refutation moves quickly, even if the result is to increase the probability of examining bad moves and decrease the probability of examining the best move. In the base position A_0, however, which necessarily lies on the critical branch, it is especially important to find the best move as soon as possible. If the tree $\mathfrak{A}$ of the contemplated game is far from uniform, an additional requirement is to give precedence to moves leading to *simple* positions, i.e. bases of subtrees of comparatively small volume.

Thus we may formulate the following requirements on the order in which positions are examined:

(1) First choose moves that have a good chance of being a best move.
(2) First choose moves that have a good chance of being a refutation, even if these have high chances of turning out to be bad.
(3) In approximately equal circumstances, prefer moves leading to subtrees of smaller dimension; for instance those in which the opponent has fewer responses.

(4) The determination of the search sequence at positions of high rank l, near the middle depth of the game tree $\mathfrak{A}$, must not be too laborious.

The first and second of these requirements conflict to a certain extent (as far as the authors know, the second is omitted in all existing game programs). At the base position A_0, and possibly in low-rank positions, the first requirement is most important. In fact, the first move (A_0, B) chosen at the base position A_0 is an improving move, even if it is rather bad. After a move satisfying the first requirement has been selected, examined, and has yielded a score, we should examine 'sharp' moves, which may turn out to be a refutation or bad. If there is reason to suppose that the opponent has already made a bad move on the branch leading to the next position A from the base position A_0, we may be guided by this requirement and choose an improving move rather than a refutation move.

Let us now take up two questions: What features of positions and moves may have a bearing on the search sequence? and how do we formulate the rules for choosing the next move in a forward step. We may use the same features that we applied in computing the value of the evaluation function $f(A)$. Suppose that the differences $f(B_i)-\mathrm{sc}(B_i)$ between the evaluation function values and the true scores, at the positions B_i reached by the moves (A, B_i) from the next position A, are independently distributed random variables with identical distribution functions. Then these moves should be considered in the order of decreasing (increasing) values of the evaluation function $f(B_i)$ if White (Black) is to move at A. In fact, in an arbitrary subset of such moves from a White position, the move to the position B_i with the highest value of the evaluation function has the highest probability of being a best or refutation move, independently of what other moves may have been examined; when A is a Black position, the minimum value of $f(B_i)$ yields the highest probability.

The assumptions justifying such an order of search are generally not satisfied for the games in which the authors (and the reader) are interested. The assumption that all the random variables $f(B_i)-\mathrm{sc}(B_i)$ are identically distributed is extremely dubious; to the contrary, there is reason to assume that one can find easily computed position features with high expected absolute values of these differences. The above method for choosing the next move suffers from two fundamental shortcomings:

(1) If the chosen move $\Psi_i=(A, B_i)$ is among the first and turns out to be a refutation move, then it is a waste of time to process the position A through all the positions $B_{i'}$ that result from the moves $\Psi_{i'}=(A, B_{i'})\in\mathfrak{A}$ and are excluded from the search after evaluation of the function $f(B_{i'})$.
(2) A move (A, B_i) with a value of $f(B_i)$ that favors one's own side has a good chance of being an improving move rather than a refutation.

For the first reason, if A is a position with high rank, it is inexpedient to order the moves Ψ_i $(i=1,2,\ldots,m)$ leading from it in an order of decreasing or increasing value of $f(B_i)$ at their destinations B_i. Instead, we may order the moves by means of an instant evaluation function $\phi(\Psi_i)$ which can be computed without knowing the positions B_i. For example, consider the game of noughts and crosses, where the aim is to place five of one's own symbols (a nought or a cross) in an unbroken line (vertical, horizontal, or diagonal) of squares on a board of large horizontal and vertical dimensions (number of squares). A move after which there arises either an *open triplet*, consisting of three of one's own pieces in a line unblocked by the opponent, or a *half-open quadruplet* consisting of four of one's own pieces in line and blocked by the opponent at one end only, is said to be *dangerous* since it may lead to a win. It is possible to decide that a proposed move has this character without actually making it. (Such a move satisfies the third requirement: the opponent must reply either by blocking the contemplated configuration or by posing an even greater threat. Obviously the number of replies that need to be considered is much smaller than in positions without threats.)

To order the moves we may use the notion of identical moves in different positions; this concept is applicable to many games. Intuitively, two moves may be considered identical if the change in the positions of the pieces is the same, without regard to the disposition of the non-moving pieces. With this concept, the number of different moves is far less than the total number of different positions. In chess there are about 10 000 different moves, compared to not less than 10^{60} different positions. Some of them, for instance the incursion of a White Rook into an uncontested square in the 7th or 8th rank, or a Black Rook into the 1st or 2nd, or the capture of a piece having a value higher than the attacker's, may be taken as having great weight, independently of the position in which they originate. Another —castling, a standard developmental move—takes on great weight in the presence of easily determined features of the base position, e.g. the presence of many pieces on their original squares.

These methods may be regarded as static in that they are based on studying a move and the positions before and after it. The study of the dynamic circumstances, connected with the search process itself, is much more important. We can gain much more useful information as a by-product of the search than we can by an over-laborious analysis of a move and the positions before and after it. To be sure, the dynamic method relates to other positions; however, neighboring positions in a search tree differ only in the locations of a few pieces, and the properties of the positions are relatively stable under such changes.

In fact, the values of the evaluation function depend linearly, or almost linearly, on features defined by the location of only a few pieces; there are many features and each piece influences a number of them. But, in positions close to each other in the game tree $\mathfrak{A}$ most of the pieces are in identical locations, and the positions all have or do not have the corresponding features. Also, the rules of the game allow many identical moves to be made in these positions, and the features change in similar ways after these moves (all the more so when the features depend on the moves and not on the positions). Finally, many features depend on the locations of slowly moving pieces—such as Pawns and Kings in chess—which if they do move do not move very far. The same property is exhibited by features that do not play the same role in ordering the search sequence as they play with respect to the value of the evaluation function, e.g. when they determine the probable accuracy of the function.

The above arguments imply that moves which are best or improving in one position have a good chance of being best or improving in neighboring positions, and that moves leading to the root of a small subtree will probably preserve this property. In the search process, therefore, it is worth while to collect statistical information on the quality of the moves and use it for choosing the next move. The portion of the program that attends to such information and to the choice of move on the basis of it, is called the *best move table* (ordinarily we do not distinguish between best and improving moves).

This routine may use an ample statistical data base, requiring a large memory, capable of holding all the desired information for say 10 000 moves in chess. The principal difficulty lies in the need to search this massive memory for the information required about each admissible move at each next position, or conversely, for many moves having relevant information stored in the memory, to determine whether they are or are not admissible in a given position. This takes a great deal of machine time and is undesirable, at least for positions of high rank in the search tree.

Normally the best move table preserves information only about moves that have been explored in the search process, producing a rough list of best and improving moves. One method is the use of the so-called substitution scheme. Suppose the program has calculated and preserved l best moves for each side. Then we have two arrays $\Xi_w[i]$ and $\Xi_b[i]$, each holding l elements. When we want to apply the deepening rule during the search and we must choose the next move from the White (Black) position A, we first look for the move $\Xi_w[i]$ ($\Xi_b[i]$) with the lowest index k, $1 \leq k \leq l$, that is admissible at A within the rules of the model game, and not yet examined. If there is no such move, we select moves in an order determined by a static process.

The arrays $\Xi_w[i]$ and $\Xi_b[i]$ are updated when we step backward. Suppose such a step is made from a White (Black) next position B. If this position is unreachable by the opponent, we do not update. Otherwise we know the best or improving move $H = (B, C)$ for Black (White) from the position C. This move may or may not be in the corresponding array $\Xi_w[i]$ ($\Xi_b[i]$). If it is, the move H is pushed upward in the array, i.e. if $H = \Xi_{w(b)}[i]$ ($i = 1, 2, \ldots, l$) we make the substitutions

$$R := \Xi_{w(b)}[i];$$
$$\Xi_{w(b)}[i-1] := \Xi_{w(b)}[i];$$
$$\Xi_{w(b)}[i] := R;$$

where R is a temporary entry used to exchange elements in the array. Obviously if $H = \Xi_w[1]$ ($\Xi_b[1]$) it is already at the top of the stack and cannot be pushed upward.

If H is a new move (perhaps a best move not preserved in memory) it is inserted in the k-th place in the array; elements with index from k to $l-1$ are pushed down and the l-th element is erased:

$$\Xi_{w(b)}[i] := \Xi_{w(b)}[i-1] \; (i = k+1, \ldots, l);$$
$$\Xi_{w(b)} := H.$$

The substitution scheme is parameterized by k and l. Given maximum confidence in the data base, $k = l$, i.e. a new move is placed at the last position; given minimal confidence, it is placed first, i.e. $k = 1$.

Maximum confidence is founded on the assumption that information in the data base about best moves is mostly reliable, i.e. consists of frequently occurring best moves, especially in positions approximating the one under examination; the appearance of a new move often turns out to be a random nuance. Minimum confidence is founded on the assumption that the set of *candidates for best move* changes rather quickly, and the appearance of a new move most often signals such a change. Both of these assumptions are extremes, and the value of k is usually set somewhere between 1 and l.

Let us now consider a somewhat more complex best move service routine, in which we take account (even though roughly) of three factors: the frequency with which the moves turn out to be best or refutations; the size of the corresponding subtrees; and the nearness of the positions in which these moves occurred to the one under examination. We are to choose the next move $H = (A, B)$, from a position A during a step forward to a position B among those accessible from A and not yet investigated. We compute the value of a *priority function* $\Psi(H)$ depending on the above parameters and on static features, after which we choose the move H having the highest priority value. (To save time, we may use a simpler method of choosing a move for positions of high rank.)

As in the simpler variant, the arrays in the tables are updated during a backward step, when the best or refutation move $H(B, C)$ from the next position B is known. We now describe the updating of the elements of the array for the corresponding color. We have the following data bases:

$\Xi[i]$—moves appearing in the table;

$\nu_w[i]$, $\nu_b[i]$—the number of cases in which the corresponding moves were best moves and refutations;

$\gamma_w[i]$, $\gamma_b[i]$—the mean number of positions, normalized with respect to the standard depth, in the corresponding subtrees of the search;

$\rho_w[i, j]$, $\rho_b[i, j]$—the shortest distances, measured in the search tree $\tilde{\mathfrak{A}}'$, between the positions A_j in the branch $W(A_0, A_1, \ldots, A_h)$ that connects the base position A_0 with the next position A_h, and the positions in which the move $\Xi[i]$ appears as a best or refutation move, respectively.

On a backward step from A_h, when the best (or refutation) move $H=(A_h, B)$ is known, the move is inserted in the array $\Xi[i]$. If it is a new move it displaces some move $\Xi[i]$ and takes its place, while we define the initial values of the parameters as:

$$\nu_w[i] := \nu_b[i] := \gamma_w[i] := \gamma_b[i] := 0;$$
$$\rho_w[i,j] := \rho_b[i_0,j] := \infty \qquad (j=h,h-2\ldots).$$

The symbol ∞ is to be interpreted as a number larger than any possible distance between nodes of the search tree $\tilde{\mathfrak{A}}'$. The displaced move must be rarely encountered and far away from the positions A_j.

Depending on whether the move $H=\Xi[i]$ is a best or refutation move we next modify the parameters with subscript b or r:

$$\gamma_{w(b)}[i] := \frac{\gamma_{w(b)}[i]\times\nu_{w(b)}[i]+\mu/M_{w(b)}(k)}{\nu_{w(b)}[i]+1};$$
$$\nu_{w(b)}[i] := \nu_{w(b)}[i]+1;$$
$$\rho_{w(b)}[i,j] := \min(\rho_{w(b)}[i,j], h-j) \qquad (j=0,1,\ldots,h);$$

Here μ is the size of the B-subtree of the search tree $\mathfrak{A}$ and $M_{w(b)}(k)$ is the mean size of the corresponding subtree at a k-th rank node among the candidates for the critical branch ($M_w(k)$) and the positions unreachable by the opponent ($M_b(k)$). These values are determined experimentally or calculated theoretically on the basis of various assumptions about the nature of the search and the tree of the model being used.

On a forward step from a k-th rank position A_k the distances $\rho_w[i,k]$ and $\rho_b[i,k]$ must be recalculated for all moves in the table $\Xi[1]$, $\Xi[2],\ldots,\Xi[l]$:

$$\rho_{w(b)}[i,k] := \rho_{w(b)}[i,k-1]+1.$$

Strictly speaking, these distances are calculated for both sides, White and Black. But in games where White and Black move alternately, the distances between positions need to be saved for one color only. Then we introduce a fictitious rank of order -1, and the distances for backward and forward steps are calculated by the formulae:

$$\rho_{w(b)}[i,j] := \min\{\rho_{w(b)}[i,j],\ h-j | j=h,h-2,\ldots,0 \text{ or } -1\};$$
$$\rho_{w(b)}[i,k] := \rho_{w(b)}[i,k-2]+2 | i=1,2,\ldots,l.$$

In the games we have studied we have found moves after which the standard best moves are unsuitable, and specifically tailored moves are required. Arguments like those given above show that the property of being a specific reply Θ to a move H must also be relatively stable. A supplementary routine for generating best replies was therefore proposed, in addition to the best move generator; this would maintain pairs of moves (H,Θ) if it turned out that the best or refutation reply to a move H from some position

A was not derived from the best move generator. There are far fewer specific best replies than best moves.

At the outset of our work, the tables of best moves (and replies) were either empty or were filled with the results of the search for the preceding move in the play (they could have been filled with some standard entries before beginning a game or, more exactly, before the first move out of the opening book, but this was not done). Since the data base was still small, these tables were badly fouled by random moves. The best move routine quickly generated moves of the required quality. A proper development of the best reply was possible only after this was done, since earlier the best and refutation moves were rarely found in the array, and moves were often random. However, when the best move generator began to work well, moves to be inserted in the best reply array were encountered only very rarely, and their updating and correction was slow. This slowness, and the rather large output of the best move generator, may explain why the best reply generator produced no significant effect.

Let us take up the question of what preliminary work can be done in low-rank positions to sharpen the order in which positions are searched. We might, for example, apply a more complex logic in the best move generator; in high-order positions we might limit ourselves to a simple displacement scheme. But principally we should look at the possibility of applying a supplementary search in low-order positions. One of the methods to be applied is the so-called *iterative search*. Rather than using a single model $\mathfrak{A}'$ of the original game tree $\mathfrak{A}$, the algorithm uses several models (usually two), which increase in volume. Normally the models are of a single type, with increasing depth of search n.

At the outset the search is carried out in a small game tree $\mathfrak{A}'$, while the search tree $\tilde{\mathfrak{A}}'$ and all the final values of the pseudoscores $\mathrm{bd}(A)$ for positions $A \in \tilde{\mathfrak{A}}'$ are stored in memory. This allows us to determine which move among the candidates for the critical branch is best, and which was the refutation move at positions unreachable by the opponent. In the search of the larger game tree $\mathfrak{A}''$ in the positions $A \in \tilde{\mathfrak{A}}'$ we first choose the moves that appear as best or refutations in the game $\mathfrak{A}'$. These will often carry over their properties to the larger model. However. if we must consider a low-ranked position $A \notin \tilde{\mathfrak{A}}'$ we apply the methods described above for choosing the move. The number of positions in the tree grows at least like $m^{n/2}$, where m is the mean number of moves (*fanout*) at the positions in the tree and n is the mean depth of the search; therefore the preliminary search of the reduced tree $\mathfrak{A}'$ has little influence on the overall search time.

But this means that it is reasonable to use more information in the preliminary search than the α, β-heuristic can provide. In particular we could find scores in the search at small depths for all positions B_i after the moves (A_0, B_i) $(i = 1, 2, \ldots, \mu)$ from the base position A_0. The α, β-heuristic needs to be applied only for positions of rank higher than 1. However, a different method seems better for extending the search tree and enlarging the information base at low-rank positions. We can weaken the pruning rule so that it applies only when the pseudoscores for the next position, as defined by the positions already examined, significantly exceed the attainable bounds:

$$\operatorname{bd}(A_k) \geq \overline{\lim} + \delta = \max\{\operatorname{Bd}(A_i) + \delta | A_i \in \Pi_b\} \text{ if } A_k \text{ is a White position,}$$

$$\operatorname{bd}(A_k) \leq \underline{\lim} - \delta = \min\{\operatorname{Bd}(A_i) - \delta | A_i \in \Pi_w\} \text{ if } A_k \text{ is a Black position,}$$

where A_k is the next position, of rank k, and Π_w, Π_b are respectively the sets of those positions in the branch $(A_0, A_1, \ldots, A_k)$, leading from the base position A_0 to the next position A_k, for which the next moves are White and Black; $\delta > 0$ is chosen so that we investigate moves of sufficiently high quality.

Finally, we try to extract from the search of the model tree $\mathfrak{A}'$ enough information to prescribe the next approximation to the original game—the model $\mathfrak{A}''$. But this is not related to the determination of the order in which the search is conducted. We note only that if we follow this path we had best develop our model game in stepwise fashion by deciding at each step which position is to be added to our existing model tree.

In many games we find equivalent positions, which arise at different places in the game tree. For instance, in games where pieces move around on a board, equivalent positions occur with the same collocations of the pieces and the same side having the move (in chess, we must also know whether the position has arisen earlier, and in some positions whether the right to castle in one direction or the other has been lost, and whether capture *en passant* is permissible). A study of the game of *odnomastka* (the name means 'single-suit') will be found in [11]; this is played by algorithms rather than by people (it is a simplified version of such human games as Boston or bridge). A stronger definition of equivalence of positions can be given for this game.

In odnomastka there are $2s$ pieces, ordered by strength, which are dealt equally by some process to two players. The play consists of a sequence of cycles; in each cycle one player selects one of his pieces, and the second player does the same thing; the selected pieces are compared, the player with the stronger piece scores one point, and the two pieces are discarded. White chooses first on the first cycle; afterward, the scoring player chooses first. The goal of the game is to get the highest point score. After each cycle the number of pieces in the hand of each player is reduced by 1. A position may be *compressed* by discarding from the ordered sequences belonging to both players those pieces that will not be played in the future. Clearly, positions with the same configurations of pieces after such compression, and with the same turn to move, are equivalent.

To shorten the search it may be useful to know whether a position equivalent to the next one has been met before, and if so, what can be said about its score. If we merely preserve all the previously encountered positions and information about their scores, in the form of a single table, or as a memory of the structure of the tree, we will spend as much time on the average in finding an equivalent position as we would spend in searching the subtree of the given position. Some special table structures have been developed that allow a given element to be found quickly, without searching the entire table. For these, the table is said to be *dynamic*. Its makeup varies in the course of its use. Several papers have been devoted to the development of effective structures for dynamic tables, e.g. [3, 17, 24].

For the study of games with large but searchable numbers of non-equivalent positions an effective search can be made upward from the terminal positions rather than downward from the base position. This is the way in which chess endgames with five to ten pieces in play have been investigated: King, Queen, and Pawn against King and Queen; or King, Rook, Pawn against King and Rook (see [14], [21]). In such endgames, and those that can be derived from them by promotion of Pawn and captures of pieces, there are more than 10^9 non-equivalent positions, and in searching the corresponding subtrees a substantially greater number of nodes must be examined.

For simplicity we shall assume that in the games we now contemplate White and Black will move alternately and that there are only two outcomes, e.g. a win by the stronger side, and the impossibility of such a win. As in problems and studies, we shall assume that White is the stronger side. Positions won for White may be classified by rank, and we shall denote by $R[i]\|(i=0,1\ldots)$ the sets of positions won in i moves. The set $R[0]$ can be determined by a single inspection of all Black positions to see whether they are terminal and what their score is. In fact, $R[0]$ contains all the positions at which we can determine either immediately or after a short search that they are won for White; for instance, positions in which White can promote a Pawn to Queen or Rook and Black has no stalemate nor can check or immediately capture a White piece.

The basic tool for the investigation is a *backward step* to a position from which a legal move can be made to the given position. Suppose that we have already constructed certain sets $R[0]$, $R[1], \ldots, R[k]$ of positions with rank not exceeding k and $M[i]\|i=1,2,\ldots,k$ of positions not yet ordered and belonging to the corresponding color:

$$M(2i) = M_b \setminus \bigcup_{j=0}^{i} R[2j],$$

$$M[2i+1] = M_w \setminus \bigcup_{j=0}^{i} R[2j+1] \| i=1,2,\ldots,\ulcorner k/2 \urcorner,$$

where M_w and M_b are the sets of all mutually non-equivalent positions in which the corresponding color has the move. We denote by U' the set of positions that can be obtained from the positions in U by a backward step, i.e. those from which a move admissible in our game will arrive at a position in U. From any position in the set $R[2i+1]$ we may legally move to a position in $R[2i]$, and any position from which we may make such a move is won in no more than $2i+1$ moves. Therefore

$$R[2i+1] := R'[2i] \setminus \left(\bigcup_{j=0}^{i=1} R[2j+1] \right)$$
$$:= R'[2i] \cap M[2i-1].$$

In the same way we prove that the set $R[2i+2]$ consists of positions in the set $M[2i]$ from which all moves lead to positions of lower rank, at least one of them leading to a position of rank $2i+1$. But this means that it contains those and only those positions in $M[2i]$ to which no backward step can be made from positions in $M[2i+1]$:

$$R[2i+2] := M[2i] \setminus M'[2i+1];$$

moreover

$$M[i+1] := M[i-1] \setminus R[i+1] \ (i=0,1,\ldots).$$

We construct the sets $R[i]$ and $M[i]$ one after the other. Since for different i the sets $R[i]$ do not intersect, and the total number of positions is finite, the set $R[i]$ is empty for some i and $M[i]=M[i-2]$. Then no win for White is possible at positions in the sets $M[i]$ and $M[i-1]$.

The implementation of this algorithm for studying a game suffers from specific difficulties stemming from the fact that the sets $R[i]$ and $M[i]$ are not wholly stored in fast memory. The methods of coping with the corresponding problems are unrelated to the theme of this book.

The Construction of Models of a Game

Suppose given two game trees $\mathfrak{A}$ and $\mathfrak{A}'$ in which every move (A'_0, B') from the base position (A'_0) in $\mathfrak{A}'$ is mapped into some corresponding move (A_0, B) from the base position A_0 in $\mathfrak{A}$. Then we may choose a move from A_0 by finding a best move (A'_0, B') from the base position A'_0 in $\mathfrak{A}'$ and then choosing its image (A_0, B) in $\mathfrak{A}$. This process allows us to examine fewer positions if $\mathfrak{A}'$ is significantly smaller than $\mathfrak{A}$, and it is meaningful when we have some reason to expect that a best move in $\mathfrak{A}'$ maps into a good move in $\mathfrak{A}$.

Let us consider some techniques that, given a tree $\mathfrak{A}$, will construct a significantly smaller tree $\mathfrak{A}'$ which will to some degree satisfy our desire that its best moves map into good moves. Starting from the root A_0 of $\mathfrak{A}$, we add, one after another in $\mathfrak{A}'$, the moves belonging to $\mathfrak{A}$ from the positions already in $\mathfrak{A}'$, together with the positions to which they lead. At the newly added position the turn to move, and the score if the new position is terminal in $\mathfrak{A}$, are the same as they are in $\mathfrak{A}$. When no move $(A, B) \in \mathfrak{A}$ from a non-terminal position $A \in \mathfrak{A}'$ is contained in $\mathfrak{A}'$, we use some definite method to compute a score $\mathrm{sc}(A)$ that need not coincide with the score of this position in the original game.

Further, at some positions $A \in \mathfrak{A}'$ the side that has the move is allowed a choice: to make one of the moves $(A, B) \in \mathfrak{A}$ that are permissible in $\mathfrak{A}'$, or to accept a score (usually computed as for a non-terminal position in $\mathfrak{A}$) and take the position as terminal in $\mathfrak{A}'$. The possibility of scoring the position $A \in \mathfrak{A}$ is assumed to be due to a *blank move* $(A, A') \in \mathfrak{A}'$. Blank moves lead to terminal positions in $\mathfrak{A}'$ that are not found in $\mathfrak{A}$, and from a position in $\mathfrak{A}'$ only one blank move may be made. Since some move in $\mathfrak{A}$ must correspond to a move $(A_0, B) \in \mathfrak{A}'$, it is impossible to make a blank move from the base position $A_0 \in \mathfrak{A}'$.

The tree $\mathfrak{A}'$ that we construct in this way will be called a *model of the game tree* $\mathfrak{A}$. The construction process can be combined with a search. At each step in the search we may add the following modelling actions to those described in Chapter 1:

(1) deciding which moves $(A, B) \in \mathfrak{A}$ from the next position are admissible in $\mathfrak{A}'$:
(2) deciding whether a blank move (A, A') is admissible;
(3) computing the score at a position A that is non-terminal in $\mathfrak{A}$ and terminal in $\mathfrak{A}'$;
(4) computing the score at a position $A' \notin \mathfrak{A}$ arising from a blank move $(A, A') \in \mathfrak{A}$.

We shall ignore the problems of computing the scores, although we have said something about them in the first section of this chapter. We merely note that we take account of the features of the position $A \in \mathfrak{A}$ when we compute the score of a position A' arising from a blank move $(A, A') \in \mathfrak{A}$.

A model $\mathfrak{A}'$ of a game tree $\mathfrak{A}$ is said to be *equivalent* if it satisfies the following two conditions:

at every non-terminal position $A \in \mathfrak{A}'$ a non-blank best move (A, B) in $\mathfrak{A}'$ is also a best move in $\mathfrak{A}$;

when a blank move (A, A') in the game $\mathfrak{A}'$ is a best move, there exists a best move (A, B) in $\mathfrak{A}$ from the position A to a position $B \notin \mathfrak{A}'$.

For any game tree $\mathfrak{A}$ containing more than two positions we can construct a smaller equivalent model. Let $\mathfrak{A}'$ be a subtree of $\mathfrak{A}$ having its root at the base position A_0 and containing the first-rank position B such that (A_0, B) is a best move in $\mathfrak{A}$. It may be looked on as a game tree with scores at its terminal positions equal to their scores in $\mathfrak{A}$, and is equivalent to $\mathfrak{A}$. However, to make use of such a model, we need a method for determining the scores at its terminal positions.

In some cases we can find these scores without searching too many superfluous positions in the original tree $\mathfrak{A}$. In an endgame with King, Bishop, and Pawns against King and Bishop we may neglect variations in which the weaker side sacrifices a Bishop for a Pawn. In fact, there are no positions in an endgame with King and Bishop against a lone King in which either side can mate, and so all positions in such an endgame are drawn. Later we shall examine several somewhat more general methods for constructing equivalent models but, as we showed in Chapter 1, any such method must be founded on various properties of the rules of the game we are studying.

Game-playing algorithms often use models whose equivalence to the original game is either not proven or non-existent, i.e. they use heuristic models, whose plausibility is usually based on approximations. For some of these models we shall show later how we can compute the probabilities of correctly choosing a best move at positions in the original game $\mathfrak{A}$ or of computing their scores, whenever some definite information about the positions can be obtained without searching the corresponding subtree of the game $\mathfrak{A}$. Here we describe these models and the properties of the games that allow their use.

First we take up what we may call the *formal* methods of building models that require the least amount of analysis of the contemplated positions and moves. We may, for instance, simply set an *a priori* limit to the number of moves in $\mathfrak{A}$ that will be examined at the positions in the game $\mathfrak{A}'$. In the simplest case the limit depends only on the rank of the positions. We prescribe a function $\varphi(k)$ of the integer argument k, usually monotone decreasing and vanishing for some $k = n$. All moves (A, B_i) from White (Black) positions $A \in \mathfrak{A}'$ are arranged in order of decreasing (increasing) values of some function $f_{\text{ord}}(B_i)$ and we admit to the game $\mathfrak{A}'$ those and only those moves (A, B_i) for which $i \leq \varphi(k)$, where k is the rank of the position A.

We often use for the move-ordering function $f_{\text{ord}}(B)$ a function $f_{\text{sc}}(B)$ which determines the scores of the terminal positions in the game $\mathfrak{A}'$. In fact, however, these functions are subjected to variegated conditions. The function $f_{\text{sc}}(A)$ must yield values as close as possible to the true scores $\text{sc}(A)$ of the positions in the original game $\mathfrak{A}$. The values of $f_{\text{ord}}(A)$ should be such that the number of best moves (A, B_i) from positions $A \in \mathfrak{A}'$ has a high probability of being less than the prescribed limit. As we noted in Section 2, this implies that low numbers should be assigned not only to moves that lead to positions with extremal values of the evaluation function $f_{\text{sc}}(B_i)$ (maximal for White positions and minimal for Black) but also to moves leading to large expected values of the difference $|\text{sc}(B_i) - f_{\text{sc}}(B_i)|$.

Let $\mathfrak{A}''$ be a Shannon model of depth n of the same game $\mathfrak{A}$ with identical values of the evaluation function $f_{\text{sc}}(A)$ at its terminal positions. Then if we already have $\varphi(n) = 0$, the tree $\mathfrak{A}'$ constructed according to the above rules is a model of the game $\mathfrak{A}''$. If $\varphi(n-1) > 0$ and if at every non-terminal position $A \in \mathfrak{A}'$ at least one move (A, B) which is a best move in $\mathfrak{A}''$ is admissible in $\mathfrak{A}'$, then $\mathfrak{A}'$ is an equivalent model of $\mathfrak{A}''$ and the scores of all positions $A \in \mathfrak{A}'$ are the same in both games. When the best moves in $\mathfrak{A}''$ are not necessarily admissible in $\mathfrak{A}'$ we may ask about the probability of correctly estimating the score and choosing the best move by using the model $\mathfrak{A}'$.

We can estimate these probabilities for a uniform game $\mathfrak{A}_{m,n,s}$ of the type introduced above, having a tree depth n, alternate moves by White and Black, fanout m at positions A of rank $k < n$, s winning moves at positions where a win is possible, and scores $f_{sc}(A)$ having the values 0 or 1 at the terminal positions, all of which have rank n; all this provided we assume mutual independence among the probabilities that moves $(A, B) \in \mathfrak{A}_{m,n,s}$ are admissible in the model, and assume that for winning moves (A, B) these probabilities depend only on the rank k of the position A at which they can be made:

$$P\left((A,B) \in \mathfrak{U}' | sc(A) = sc(B) = \begin{cases} 1, & \text{if } A \text{ is a White position} \\ 0, & \text{if } A \text{ is a Black position} \end{cases}\right)$$
$$= \pi_k (0 \le k \le n)$$

Let p_k and q_k be the respective probabilities of obtaining a correct score for positions of rank k when we can or cannot make a winning move from them. The terminal positions will be correctly scored, i.e. $p_k = q_k = 1$. All moves from a losing position A lead to positions at which the opponent has a win. For the score at A to be correct, the scores of positions that can be reached from it by admissible moves in the game $\mathfrak{A}'$ must also be correct. Since the corresponding subtrees of $\mathfrak{A}'$ do not intersect, the probabilities of these events are independent, and we have the recursive relationships

$$q_k = p_{k+1}^{\phi(k)}, \qquad 0 \le k < n.$$

Let there be t winning and $\varphi(k) - t$ losing moves among the admissible moves from some winning position A. Then in order that the score at A be wrongly computed, the scores after winning moves must also be wrongly computed and after losing moves correctly computed. The probability of this event is equal to $(1 - q_{k+1})^t p_{k+1}^{\varphi(k)-t}$, where k is the rank of the position A. The probability that the contemplated case will occur is $C_s^t \pi_k^t (1 - \pi_k)^{s-t}$. Consequently the total probability of error $1 - p_k$ is equal to

$$\sum_{t=0}^{\min(s, \varphi(k)-s)} C_s^t \pi_k^t (1 - \pi_k)^{s-t} (1 - q_{k+1})^t p_{k+1}^{\varphi(k)-t}$$

and for $s \le \varphi(k) - s$

$$p_k = 1 - \left(\pi_k (1 - q_{k+1}) + (1 - \pi_k) p_{k+1}\right)^s p_{k+1}^{\varphi(k)-s}$$
$$(0 \le k < n).$$

Table 3. Probability that Scores Will Coincide When Some Moves are Not Considered

Notation

m—number of moves at non-terminal positions in the model,
s—number of winning moves at winning non-terminal positions,
n—depth of the uniform game,
π—probability that a winning move will have number $v \leq m$
P_n—probability of a win in the model game at a won position in the original game,
Q_n—probability of a win in the model game at a lost position in the original game.

	$s=2,\ m=3,\ \pi=0.8$		$s=2,\ m=5,\ \pi=0.8$		$s=2,\ m=3,\ \pi=0.9$	
n	P_n	Q_n	P_n	Q_n	P_n	Q_n
0	1.0000	0.0000	1.0000	0.0000	1.0000	0.0000
1	0.9600	0.0000	0.9600	0.0000	0.9900	0.0000
2	0.9646	0.1153	0.9674	0.1846	0.9903	0.0297
3	0.9216	0.1025	0.8946	0.1528	0.9843	0.0288
4	0.9347	0.2173	0.9350	0.4270	0.9848	0.0463
5	0.8783	0.1835	0.7716	0.2852	0.9807	0.0450
6	0.9087	0.3224	0.9328	0.7265	0.9812	0.0569
7	0.8244	0.2497	0.5215	0.2937	0.9781	0.0554
8	0.8904	0.4397	0.9837	0.9614	0.9786	0.0642
9	0.7500	0.2942	0.1120	0.0790	0.9763	0.0627
10	0.8889	0.5781	0.99999	0.99998	0.9768	0.0694

	$s=2,\ m=5,\ \pi=0.9$		$s=2,\ m=5,\ \pi=0.95$		$s=2,\ m=5,\ \pi=0.99$	
n	P_n	Q_n	P_n	Q_n	P_n	Q_n
0	1.0000	0.0000	1.0000	0.0000	1.0000	0.00000
1	0.9900	0.0000	0.9975	0.0000	0.9999	0.00000
2	0.9905	0.0490	0.9975	0.0124	0.9999	0.00050
3	0.9801	0.0466	0.9962	0.0123	0.9999	0.00050
4	0.9815	0.0957	0.9963	0.0187	0.9999	0.00055
5	0.9679	0.0889	0.9955	0.0185	0.9999	0.00055
6	0.9717	0.1506	0.9955	0.0224	0.9999	0.00056
7	0.9503	0.1339	0.9950	0.0222	0.9999	0.00056
8	0.9601	0.2249	0.9951	0.0247	0.9999	0.00056
9	0.9212	0.1841	0.9947	0.0245	0.9999	0.00056
10	0.9480	0.3366	0.9948	0.0261	0.9999	0.00056

Table 3 displays the values of p_k and $1-q_k$ for various values of k, s, and φ when the values of the $\varphi(k)$ are identical for all non-terminal positions, i.e. $\varphi(0)=\varphi(1)=\ldots=\varphi(n-1)=\varphi$. Inspection of the table shows that except when π_k is very close to 1 the model should be used only for small depths n. If the game we are studying is nearly uniform, and the probabilities that moves will be admissible are nearly independent, it is inexpedient to increase the depth of the search on account of the uniform bound on the number of admissible moves. In fact, when the best move (A, B) at some position A is not admissible, it is highly probable that at nearby positions the best moves are also inadmissible. On this account the probability of erroneous scores only increases, and it is not surprising that programs with models of the type we have described play weakly.

The bounds $\varphi(A)$ on the number of admissible moves in the model $\mathfrak{A}'$ need not depend solely on the ranks of the corresponding positions A. However, up to now no such models have been studied theoretically nor have they been used in practice for game-playing programs. We shall look at a very simple case where:

(a) in all non-terminal positions A of the game $\mathfrak{A}$ White has the move and the moves from A are ordered, while the probabilities $P(l)$ that the move $(A, B_l) \in \mathfrak{A}$ with index l is a best move are identical and mutually independent;
(b) the positions of rank n, and only those, are terminal.

Thus, we are to find the critical path $(A_0, A_1, \ldots, A_n)$ leading from the base position A_0 to the terminal position A_n having the maximum score $\mathrm{sc}(A_n)$.

Every position A_k of rank $k = 1, 2, \ldots, n$ can be defined by k coordinates $x_1, x_2, \ldots, x_k$, where x_i is the number of moves from A_i in the branch $(A_0, A_1, \ldots, A_k)$ leading to A_k from the base position A_0. Under our assumptions this branch has the probability $P(x_1) \times P(x_2) \times \cdots P(x_n)$ of being the beginning of the critical branch. Since the number of positions in the game tree $\mathfrak{A}'$ is bounded, we should first of all look at the terminal positions which have the greatest probability of being optimal. Accordingly, the game tree $\mathfrak{A}'$ must be constructed of branches leading to terminal positions having coordinates that satisfy the condition

$$P(x_1) \times P(x_2) \times \cdots \times P(x_n) \geq L,$$

where L is a suitably chosen constant.

We may often suppose that for the interesting values of x the function $P(x)$ is approximately linear, and that the linear function is nearly exponential, i.e.

$$P(x) \approx q - rx \approx qe^{-rx/q}.$$

Then the tree of the game $\mathfrak{A}'$ contains terminal positions with coordinates that satisfy the condition

$$P(x_1) \otimes P(x_2) \otimes \cdots \otimes P(x_n) \approx q^n e^{-\frac{r}{q}(x_1 + x_2 + \cdots + x_n)} \leq L,$$

i.e. $x_1 + x_2 + \cdots + x_n \leq \ln \dfrac{L}{q^n} = C$

If we begin the numbering of the moves from non-terminal positions $A \in \mathfrak{A}'$ with 0, the condition for including the positions A_k of ranks $k=1,2,\ldots,n$ in $\mathfrak{A}'$ may be stated as the inequality

$$x_1 + x_2 + \cdots + x_k \leq C,$$

and we may specify the constant C immediately, without defining r, q, and L.

When we limit the number of admissible moves in a model of a two-person game $\mathfrak{A}$, the choice of the best move may be in error because the move may be inadmissible in the model $\mathfrak{A}'$, but also for other reasons. A bad move may be preferred to a best move because its refutation is inadmissible in $\mathfrak{A}'$. Since for a move (A, B_i) with a large value of i the chance of turning out to be bad is high, we should prolong the search for a refutation as far as possible unless it comes up among the earlier responses. In a model $\mathfrak{A}'$ of a game $\mathfrak{A}$ with alternating moves by White and Black, we might postulate, by analogy with the above argument, that the condition for including the positions A_k of ranks $k=1,2,\ldots,n$ is given by the inequality

$$\Phi(A_k) = x_k - \Phi(\Phi(A_{k-1})) \leq C,$$

where (A_{k-1}, A_k) is a move in the game $\mathfrak{A}$ leading to the position A_k and $\Phi(x)$ is a monotone increasing function of x. We have not, however, succeeded in formulating a probabilistic hypothesis that would support the use of this condition.

Let us now turn to the semantic method of constructing models, in which the admissibility of moves $(A, B) \in \mathfrak{A}$ is determined on the basis of a qualitative analysis of the moves or of the corresponding positions A and B. The notion of an *unstable* position makes sense in many games. In such a position there is a rather high probability that the value of the evaluation function $f(A)$ will differ significantly from the true score $\mathrm{sc}(A)$. Moreover, either few moves $(A, B) \in \mathfrak{A}$ lead from it or the majority of them lead to bad positions. By analyzing some easily computed features of the position and of the moves leading to or from it that are admissible under the rules of the game $\mathfrak{A}$, we can determine whether or not the position is unstable.

For instance, in the game of noughts and crosses, those configurations consisting of m successive squares arranged in vertical, horizontal, or diagonal lines and occupied by one of the players with noughts or crosses, and having a continuation of 5-m contiguous free squares on one end or on both ends are called respectively half-open and open m-tuples. If the opponent has an open triplet all moves will lose except those that either immediately adjoin the triplet or make a threat to complete a quintuple of one's own. Thus positions in which the opponent has an open triplet are unstable.

In draughts—Russian, American, or international—positions in which a capture is possible are unstable. In fact, the material score in such positions may change in ways that cannot be predicted without analysis of the corresponding variations. (Moreover, in draughts captures are obligatory, so that the number of admissible moves in such positions is significantly less than in other positions.)

Chess positions arising after a capture may be considered to be unstable. As a rule, one may win back the value of a captured piece or even capture one of higher value. If one abstains from the capture, one cannot in general return to the former material balance at a later date. Thus the material score in such positions is unreliable and there are few non-losing moves. There is a substantial difference, however, between this example and the first two. In those, few of the moves from an unstable position are permitted by the rules of the game or fail to lead to obviously losing positions. In the chess example, however, it is merely highly probable that a 'quiet' move from an unstable position will be of poor quality.

Moves leading to an unstable position will be called *forcing* moves, and moves from an unstable position, except those leading to clearly or very probably lost positions, will be called *forced* moves. As is clear from the above examples, the features of forcing and forced moves are also easily computed (in game-playing algorithms for noughts and crosses, draughts, and chess we divide the unstable positions, forcing moves, and forced moves into different classes, but they all have easily computed features).

We can now define the concept of a *forced game*. Its positions are positions in the game $\mathfrak{A}$ and the admissible moves are forcing, forced, and blank moves. The admissibility of blank moves is related to the fact that in the games we are considering almost every position can be mapped into a position with the same configuration of the pieces, but with the move to be made by the side with the opposite color. If no such position exists (as in chess with a King in check), a blank move is inadmissible. Sometimes, however, a blank move is disallowed in other positions, as when a player under the threat of material loss will make an intermediate checking move (in models of chess a check is often regarded as a forcing move) against which his opponent defends himself; if a blank move were permitted, the program would not perceive a threat of material loss.

The notion of the forced game can be used to construct a model in the following way. As in Shannon's model the depth of search n is prescribed. In positions A_k of rank $k \leq n$ all moves in the game tree $\mathfrak{A}$ are admissible, but in contrast to Shannon's model, positions of rank n are taken to be non-terminal (unless they are terminal in $\mathfrak{A}$). At these and at all positions of higher rank, every move in the forced game is admissible. The ranks of positions in the model $\mathfrak{A}'$ may be either unbounded, or bounded by a search depth $n_1 \gg n$.

In many cases we can construct a model $\mathfrak{A}''$ equivalent to the model $\mathfrak{A}'$ described above but having fewer moves. To simplify the description of such a model we shall assume that: 1) in the original game $\mathfrak{A}$ White and Black move alternately; 2) the evaluation function $f(A)$ is the sum of material and positional components

$$f(A) = f_{\mathrm{m}}(A) + f_{\mathrm{p}}(A).$$

The material scoring function is linear:

$$f_{\mathrm{m}}(A) = \sum_{\mu=1}^{M} h_\mu P_\mu^w(A) - \sum_{\mu=1}^{M} h_\mu P_\mu^b(A),$$

where h_μ is the weight of the μ-th piece, M is the number of different pieces belonging to one of the sides, and $P_\mu^{w(b)}(A)$ is a predicate having the value 1 if the White (Black) piece is on the board and 0 if it is not. The values of the positional component $f_{\mathrm{p}}(A)$ of the evaluation function are non-negative and less than the minimum absolute difference $|f_{\mathrm{m}}(A') - f_{\mathrm{m}}(A'')|$ of the values of the material components of the scores.

Let A_k be the next position in the search tree being constructed for the game $\mathfrak{A}''$, with $k = n-1$ and White (Black) to move. Let $\underline{\lim}$ ($\overline{\lim}$) as defined in Chapter 1 be the bounds for the scores

$$\underline{\lim} = \min\{\mathrm{bd}(A_i) | A_i \in P_w\},$$
$$\overline{\lim} = \max\{\mathrm{bd}(A_i) | A_i \in P_b\}$$

(we recall that P_w and P_b are the sets of positions in the branch $(A_0, A_1, \ldots, A_k)$ leading from the base position A_0 to A_k, the next position for White or Black, respectively; the bd A_i are the intermediate values of the parameters of the positions in this branch as calculated by the α, β-procedure instead of their scores). After the move $(A_k, A_{k+1}) \in \mathfrak{A}''$ there arises the position A_{k+1} with Black (White) to move and with the score

$$\begin{aligned} f(A_{k+1}) &= f_{\mathrm{m}}(A_{k+1}) + f_{\mathrm{p}}(A_{k+1}) \\ &= f_{\mathrm{m}}(A_k) + (\pm) h + f_{\mathrm{p}}(A_{k+1}), \end{aligned}$$

where h is the weight of the pieces captured in the move, or is 0 if (A_k, A_{k+1}) is a quiet move. The rank of A_{k+1} is not less than n, so that in general a blank move is admissible there.

The value of $\underline{\lim}$ ($\overline{\lim}$) is equal to the value of the evaluation function for some $A \in \mathfrak{A}$ and may be represented as a sum of material and positional components:

$$\underline{\lim} = \underline{\lim}_{\mathrm{m}} + \underline{\lim}_{\mathrm{p}},$$
$$\overline{\lim} = \overline{\lim}_{\mathrm{m}} + \overline{\lim}_{\mathrm{p}}.$$

If

$$\begin{aligned} f_{\mathrm{m}}(A_{k+1}) &= f_{\mathrm{m}}(A_k) + h < \underline{\lim}_{\mathrm{m}}(f_{\mathrm{m}}(A_{k+1})) \\ &= f_{\mathrm{m}}(A_k) - h > \overline{\lim}_{\mathrm{m}}, \end{aligned}$$

then

$$\underline{\lim} - f(A_{k+1}) = \underline{\lim}_{\mathrm{m}} + \underline{\lim}_{\mathrm{p}} - f_{\mathrm{m}}(A_{k+1}) - f_{\mathrm{p}}(A_{k+1})$$
$$\geq \underline{\lim}_{\mathrm{m}} - f_{\mathrm{m}}(A_{k+1}) - f_{\mathrm{p}}(A_{k+1}) > 0,$$
$$f(A_{k+1}) - \overline{\lim} = f_{\mathrm{m}}(A_{k+1}) + f_n(A_{k+1}) - \overline{\lim}_{\mathrm{m}} - \overline{\lim}_{\mathrm{p}}$$
$$\geq f_{\mathrm{m}}(A_{k+1}) - \overline{\lim}_{\mathrm{m}} - \overline{\lim}_{\mathrm{p}} > 0,$$

since the absolute difference between the material scores of two positions is larger than the value $f_n(A)$ of the positional evaluation function for any position A.

After a blank move from the position A_{k+1} we reach a terminal position $\overline{A}_{k+1}$ in the game $\mathfrak{A}'$ described earlier, with the score

$$\mathrm{sc}_{\mathfrak{A}'}(\overline{A}_{k+1}) = f(A_{k+1}) < \underline{\lim}\ (> \overline{\lim})$$

Therefore, we also have $\mathrm{sc}_{\mathfrak{A}'}(A_{k+1}) < \underline{\lim}\ (> \overline{\lim})$, i.e. after seeing the result of a blank move from A_{k+1} we step backward to the position A_k, and the value of $\mathrm{bd}(A_k)$ remains unchanged, as though the move (A_k, A_{k+1}) had not occurred. Thus we find that at positions with rank $k \geq n-1$ and with material scores lying beyond the bounds $\underline{\lim}_{\mathrm{m}}$ or $\overline{\lim}_{\mathrm{m}}$, the only admissible moves are captures of pieces with weight $h > \underline{\lim}_{\mathrm{m}} - f_{\mathrm{m}}(A_k)(f_{\mathrm{m}}(A_k) - \overline{\lim}_{\mathrm{m}})$ or moves in the game $\mathfrak{A}'$ after which a blank reply is prohibited. The method we have just described for constructing a model $\mathfrak{A}'$ will be called an absolute scheme.

Different branches of the model $\mathfrak{A}''$ may contain different numbers of quiet moves (A_k, A_{k+1}) that are inadmissible in the forced game. In fact, for $k < n$ some moves of this kind may be non-quiet. It has been suggested that in order to improve the quality of the move (A_o, B) selected at the base position A_0 it would be useful to equalize the numbers of quiet moves in the branches, and rules for constructing models called *quiet games* have been suggested for this purpose. There are two parameters—the search depth n and the minimum number d of quiet moves in a branch. In the positions A_k of rank k any move of the original game $\mathfrak{A}$ is admissible if $k < n$ or if the number of quiet moves (A_i, A_{i+1}) in the branch $(A_0, A_1, \ldots, A_k)$ is less than d. Otherwise, only forced moves are admissible.

A program using a quiet game is stronger than one using an absolute scheme with the same search depth n and the same scoring function $f(A)$. The time spent, however, is greater. We do not yet know how to use this extra time more effectively, nor even whether we can.

Similar considerations might explain the relative success of a program playing at the expert level on a rather slow machine ([36—38]). It allows quiet moves from a position A_k under the following condition: Let A_i be a position in the path $(A_0, A_1, \ldots, A_{k-1})$ leading from the base position A_0 to the immediate predecessor A_{k-1} of A_k, and let $\psi(A_i)$ be the number of moves from A_i that are admissible in the original game $\mathfrak{A}$. Then a quiet move from A_k is allowed if the $\psi(A_i)$ $(i = 0, 1, \ldots, k-1)$ satisfy the inequality

$$\psi(A_0)\cdot\psi(A_1)\cdots\psi(A_n)\leq M,$$

where M is a parameter determining the time taken by the search from the base position A_0. In unstable positions only captures are allowed (compulsory by the rules of draughts); the number of ways to capture is much smaller than the number of quiet moves allowed in other positions. Accordingly, if many unstable positions arise on the path its length is greater than that of a quiet path, in which such positions arise infrequently.

Semantic Models

The ultimate goal of a player is to reach a terminal position with a sufficiently favorable score. During the course of the game, however, he normally tries to attain intermediate goals—to reach positions that for one reason or other please him. In the semantic theory of games, for instance chess theory, there are many features of such positions and of the moves that lead to them, but these features lack precise definitions. In game programs some such features, which are given an exact definition (usually constricting their meaning), enter into the computation of the evaluation function $f(A)$. Thus, while studying a position in the course of the game, the program attempts to reach one or another intermediate goal.

In the search process, however, many moves are examined, which are meaningless with respect to reaching the goal. For this reason there is no time for a sufficiently deep search of the sensible variations. Moreover, different favorable features of a position do not always agree. As a result, the program may pursue a middle course in a game, making a sequence of discordant moves. This is especially true of positions in which neither player threatens to reach any intermediate goal whatsoever. Then paths that lead to positions with only slightly different values of the evaluation function compete for the choice of move by the program. Often the differences are due to random nuances. The program lacks a guiding thread —it has no plan and does not hinder its opponent's plans.

A deep search is infeasible unless we abandon a study of the results of various moves from positions of low rank, but prohibiting such moves on the basis of merely quantitative scores would not lead to good results. It has therefore been suggested that, in constructing a model of a given game, we should declare moves admissible or inadmissible with the help of a semantic analysis of the moves themselves or of the positions to which they lead. An example is afforded by the classification of moves in the forced game described in the preceding section.

From the technical point of view, there are two methods for carrying out such an analysis. The first consists in studying the position A from which a move is to be made, in order to generate only admissible moves and spend no time on the others. The second method consists in studying all moves leading from A (or at least all that are not obviously inadmissible) and then making a more or less laborious analysis of each to determine its admissibility. The first method offers a potential saving in time but selects moves on the basis of very simple formal features. To decide whether to use the second method one must estimate whether less time will be spent in analyzing the admissibility of moves than would be spent in searching the portion of the tree that is excluded by that analysis from the model being constructed.

Material gain is an important and easily examined intermediate goal in a game. In particular, forcing moves often represent the threat of winning material. Some non-forcing moves containing such a threat are easily defined; for instance, attacks on undefended major pieces (these are not included in the forced game because a formal definition of a sufficiently narrow class of forced replies is difficult). A wider class of moves posing the threat of material gain can be defined with the help of an auxiliary game model.

To study the move $(A, B) \in \mathfrak{A}$ we consider the models $\mathfrak{A}_{\text{force}}(B)$ and $\mathfrak{A}_{\text{force}}(\bar{B})$ with the initial configurations of the pieces on the board the same as that in the position B. At the non-terminal positions of the two models the moves of the forced game, and only those, are admissible. The move at the base position in the game $\mathfrak{A}_{\text{force}}(B)$ belongs to the side having it at B. Thus we may assume that this base position is simply the position B. The move at the base position $\bar{B}$ of the game $\mathfrak{A}_{\text{force}}(\bar{B})$ belongs to the side having it at A. (We assume that in the original game $\mathfrak{A}$ White and Black move alternately.)

If White (Black) has the move at A, the material component of the score of $\bar{B}$ in the game $\mathfrak{A}_{\text{force}}(\bar{B})$ is compared with the material component of the bound $\underline{\text{lim}}_{\text{m}}$ ($\overline{\text{lim}}_{\text{m}}$) determined when A is the next position in the search of the fundamental model of the game $\mathfrak{A}$. If in the first case this score is not less than $\underline{\text{lim}}_{\text{m}}$, or in the second case not greater than $\overline{\text{lim}}_{\text{m}}$, the move (A, B) is said to be non-losing in the forced variation. The material component of the score of $\bar{B}$ in the game $\mathfrak{A}_{\text{force}}(\bar{B})$ may be compared with the other bounds or with the value $f_{\text{m}}(A)$ of the material scoring function at A. (It is natural to choose the latter for positions of low rank, and the former for positions of high rank.) When this score attains the corresponding bounds or lies beyond them, (A, B) is called an *active* or a *second-level forcing* move.

This model, called an *active game* scheme, has been tested in practice. In positions A_k of rank $k < n-1$ it admits all active moves that are non-losing in the forced variation. It also admits one so-called *best passive* move, which is non-losing in the forced variation but not active, and has the most favorable score (for its own side) among such moves in the game $\mathfrak{A}_{\text{force}}(\bar{B})$ (in a variant model, the game $\mathfrak{A}_{\text{force}}(B)$). If there are no active moves from

A that are non-losing in the forced variation, three best passive moves are admitted. The numbers 1 and 3 of best passive moves are program parameters. In positions A_k of rank $k \geq n$ only moves of the forced game are admitted, as in the absolute scheme.

Many decisions were taken for economy in the running time of the program. Thus the activity and safety tests of the moves (A_{n-1}, B) from positions A_{n-1} of rank $n-1$ took no less time than the search of the B-subtree consisting only of moves in the forced game, a search that might be obviated if the tests gave a negative answer (also, this search coincides with that of the tree $\mathfrak{A}_{\text{force}}(B)$ or differs from it only by somewhat wider initial bounds bd and Bd for the scores of the base position B). Therefore all moves in the original game $\mathfrak{A}$ were admissible at positions of rank $n-1$.

If the analysis of the trees in the games $\mathfrak{A}_{\text{force}}(B)$ or $\mathfrak{A}_{\text{force}}(\bar{B})$ shows that the move (A, B) lacks the necessary properties, i.e. loses in the forced game or is inactive, we may need to search the second of the two trees only to decide whether it relates to the number of best passive moves. But even this is not needed if we determine the best passive moves by comparing the scores at the base positions of whichever of the two trees was first selected for analysis. In one variant, the tree $\mathfrak{A}_{\text{force}}(B)$ is searched first and in the other $\mathfrak{A}_{\text{force}}(\bar{B})$, while the the non-selected tree is searched only if necessary. Both variants play almost identically.

In the forced game $\mathfrak{A}_{\text{force}}(\bar{A})$ the initial configuration of pieces on the board for the position $\bar{A}$ is the same as in the position A but with the move belonging to the opponent; if this game leads to a loss of material by the side having the move at A, it is highly likely that most of the quiet moves (A, B) in the game $\mathfrak{A}_{\text{force}}(B)$ will also lead to a loss of material. Clearly, such moves should first be tested for safety; then many of them will not need testing for activity. In the contrary case fewer such moves will lead to a loss (moves made from A under *zugzwang* are an exception, but this is not often found in the middle game). In this case moves should be tested first for activity and only the active and best passive moves should be tested for safety.

The rules for defining the admissibility of moves may be hybrids. For example, up to some rank n_1 all the moves in a game $\mathfrak{A}$ may be admissible, from ranks n_1 through n_2 all moves in the active game described above, and from n_2 onward only moves in the forced game. In the existing chess programs checking moves are counted as moves in the forced game only in positions up to a certain rank, or if the number of checks in the branch being examined is less than some given standard. In the program described above for the active game, all moves from the base position A_0 or from positions of rank 1 are admissible until a move is found that does not lose material when compared with the anticipated score established in an earlier search at lesser depth.

A program for the active game played rather weakly in the absolute scheme with the same depth of search n, but spent significantly less time (the variants of these programs used about the same amount of time, but were not compared). This program did not consider some moves that deserved attention, but no move included in the model tree was obviously senseless. However, variations identical from a chessplayer's viewpoint were repeatedly inspected at many positions in the search tree; their results would have been obvious to a player after they had been inspected once.

A typical example of superfluous work is the examination of so-called *pseudoactive* moves. Suppose that in the forced game $\mathfrak{A}_{\text{force}}(A)$ the opening side wins material, but does not win it in the active game following the forcing moves from A. Then many moves that have no effect on the variations in the forced game will seem to be active, but mistakenly so since after such a move (A, B) the game $\mathfrak{A}_{\text{force}}(\bar{B})$ turns out to be essentially the same as $\mathfrak{A}_{\text{force}}(A)$.

Active moves are instances of moves answering to a definite strategy, in this case the strategy of winning material. A definition of such moves is needed if we are to specify them. We might, as earlier, demand that an active move should not lose in the forced variation, but this demand seems too strong, in any case for positions of low rank. At the same time, it is scarcely worth while to call a move active if it puts a piece of its own in peril and simultaneously threatens to win material with the aid of the piece itself. Thus we must compare moves from different positions in the trees $\mathfrak{A}_{\text{force}}(\bar{B})$ and $\mathfrak{A}_{\text{force}}(B)$ among themselves.

To refine the notion of an active move we may use the fact that for many games we can define the concept of identical moves from different positions. After such a move, in all the positions where it is possible to make it, the placement of the pieces changes in the same way. An exact definition of sameness will be given in the next chapter. Here an intuitive notion will do. We may call a move (A, B) inactive if the opponent wins by the move (B, D) in the game $\mathfrak{A}_{\text{force}}(B)$ and the winning reply $(\bar{B}, C)$ is impossible in the game $\mathfrak{A}_{\text{force}}(\bar{B})$.

If a player in one of the variations has already won material it is useful for him to try defensive moves, after which his opponent cannot secure material equality. Defensive moves must be found also when the moves found earlier all lead to a loss of material. One test of the defensive qualities of a move (A, B) from the position A is the forced game $\mathfrak{A}_{\text{force}}(B)$ defined above. It may turn out, however, that a non-losing move in the forced variation will lead to a loss in the original game, which is richer in moves. A more precise definition of the conditions for a move to be defensive requires the use of concepts developed in the next chapter; nevertheless something can still be said here.

A threat of material loss is determined in some game tree, e.g. the B'-subtree $\mathfrak{A}'(B')$ of the contemplated model $\mathfrak{A}'$ of the original game $\mathfrak{A}$. Its base is at the position B', to which the preceding move (A, B') leads from the position A, the next position in the current state of the search. We ask what pieces were in fact lost in the game $\mathfrak{A}'(B')$. Sometimes it is easy to answer, as when the opponent has captured a piece standing at the instant of capture on the same square it occupied in the position A, while we could not reply by a capture of equal value. Then we could say that defensive moves would consist in retreating the piece to another square, protecting it by another piece, or blocking the path of the opponent.

The determination of intermediate goals other than the gain of material depends more substantially on the specifics of the game in question. Later we shall give an example of a certain class of model games in chess that offer a wider set of intermediate goals. These goals are the occupation of certain squares on the board by pieces in a given subset of one's own pieces, or attacks on these squares.To every intermediate goal there correspond strategies for reaching it, and to every strategy there correspond moves that answer to it. Any given move may answer to several strategies, and various strategies may be interconnected.

Suppose that to every chess piece there corresponds a board representing its potential—a 64-bit array representing the squares on the board. For those squares attacked by the piece and not occupied by pieces of its own color (these are the squares to which it can move), the corresponding bit is set to the value 1, and the remaining bits are set to 0. There are two boards per Pawn, one representing squares to which it can move, the other squares it can attack. Boards of this kind can specify the configurations of all pieces, of pieces of one color, pieces of a given kind, etc. The problems of setting up such boards and holding them in memory belong to programming methodology and are not of interest in this book.

To every elementary strategy there corresponds a square on the board which we call a null-rank square. The goal of a strategy is to to occupy the null-rank square with some (not arbitrary) piece of its own side or to attack it. Squares from which pieces can attack (for Pawns attack or move to) null-rank squares are called first-rank squares. Thus there will be first-rank squares for King, Queen, Rook, Bishop, Knight, and Black and White Pawns. Second-rank squares for a piece are those from which it can attack at least one first-rank square, not necessarily one of its own color, and so on through the higher ranks.

The general principle for defining moves responding to a given elementary strategy is this: the piece to be moved should go to a square of lower rank, or if the piece stands still the rank of the square on which it remains should decrease, else the rank of a square attacked by the opponent should increase. On the one hand, these requirements should be relaxed to permit a shift of the attack on a square of a given rank to another square of the same rank. On the other hand, they should be tightened to exclude commotions on the distant approaches to null-rank squares and to decrease the general number of strategic moves.

Let us look at a concrete example of a system for defining strategic moves. This system consists of a set of predicates $P(A, B)$ depending on the beginning and ending positions A and B of the move (A, B), or in some cases on the move itself. The move responds to the strategy if at least one of the predicates has the value 1 (is true). To widen the set of strategic moves we add to the original set one or more supplementary predicates of this kind. To narrow the set of strategic moves we can remove one of the predicates, or change it by imposing additional conditions on its truth value. In this way we can stipulate various means for representing the system in the course of its experimental trials.

First of all, moves of one's own pieces to chosen null-rank squares respond to elementary strategies. For instance, the strategy of the weak point requires the occupation of the point by minor pieces (Bishops or Knights); the strategy of the open file requires the incursion of major pieces (Rooks or Queens) into a square in the 7th or 8th rank of the file (with respect to one's own side); the strategy of the Pawn advance requires the movement of a Pawn to a null-rank square. Besides these, moves that result in new attacks on null-rank squares or on the opponent's pieces that are fighting on null-rank squares, or that result in the disappearance of the opponent's attack on null-rank squares or on one's own pieces standing on null-rank squares, are all responsive to an elementary strategy.

These features of strategic moves are easily defined with the aid of some redundant information about the positions, moves, and strategies. The positions are defined by two sets of boards: a) $\Phi_m[\mu]$ and $\Phi_e[\mu]$ for the positions of one's own and the opposing pieces; the bits are set to 1 for pieces of given types: Pawns for $\mu=1$, Knights for $\mu=2$, Bishops for $\mu=3$, Rooks for $\mu=4$, Queen (or Queens if there has been a promotion) for $\mu=5$, and King for $\mu=6$; and b) $M[i]$ $(i=1,2,\ldots,q_m)$ and $E[i]$ $(i=1,2,\ldots,q_e)$, where the bits are set to 1 for squares attacked by pieces of the corresponding type, one's own or the opponent's, and q_m and q_e are the respective numbers of these pieces. We shall examine these boards for the positions before and after a move. The first set will be marked with a prime, the second with a double prime.

A move will be described by 1) a six-bit array N in which a single bit is set to 1 in the place corresponding to the piece-type, 2) the board G' specifying the square from which the piece is moved, and 3) the board G'' specifying the destination square. Thus for a Pawn move $N=(000001)$, for a Knight move $N=(000010)$, etc. The move Bc1–g5, for instance, is described by $N=(000100)$, the board G' on which the square c1 is marked, and the board G'' on which the square g5 is marked. The strategy is described by 1) the board R_0 for the null-rank square, 2) the two boards $A_m[\mu]$, $A_e[\mu]$ specifying the squares from which, in the given position, the selected null-rank square is attackable by one's own or hostile pieces, respectively, (clearly $A_m[\mu]=A_e[\mu]$ for $\mu=2,3,4,5,6$ but $A_m[1]\neq A_e[1]$), and by 3) the six-bit array S in which a single bit is set to 1 to specify the type of the piece trying to reach the target null-rank square.

We shall perform some set-theoretical operations on these boards: a bit in the board $P\cup Q$ is set to 1 if it is set to 1 in either P or Q or both; in

$P \cap Q$ if it is set to 1 in both P and Q; in $P \setminus Q$ if it is set to 1 in P but not in Q. The predicate $[P]$ is true if and only if P is not empty, i.e. at least one bit in it is set to 1; an empty board will be denoted by $\varnothing$. We use a similar notation for operations on sets with fixed numbers of bits and on their associated predicates.

With this notation, the predicate expressing occupancy of the null-rank square by the corresponding piece is

$$[R_0 \cap G'']\&[N \cap S],$$

The predicate expressing the appearance of new attacks by our pieces is

$$\left[\left(R_0 \cup \bigcup_{\mu=1}^{6} A_e''[\mu] \cap \Phi_e''[\mu]\right) \cap \bigcup_{i=1}^{q_m} (M''[i] \setminus M'[i])\right],$$

and that for the disappearance of former attacks by hostile pieces is

$$\left[\left(R_0 \cup \bigcup_{\mu=1}^{6} A_m''[\mu] \cap \Phi_m''[\mu]\right) \cap \bigcup_{i=1}^{q_e} (E'[i] \setminus E''[i])\right],$$

where $E''[i] = \varnothing$ if the opponent's i-th piece has been captured during the move in question.

Elementary strategies are compounded into non-elementary strategies by forming the union of their null-rank squares. First we define *active* strategies, aimed at reaching some intermediate target. For the strategy of White's center (we might also call it the development strategy) the null-rank squares are e4, e5, e6, d4, d5, and d6; for Black's development strategy they are e5,e4,e3, d5, d4, d3. For an attack on the opposing King the null-rank squares are the one on which he stands and the squares that he attacks. The strategy of the open file, mentioned above, is non-elementary, and so in essence are the strategies of Pawn advance, even though at any instant they have only one null-rank square. To every active strategy there corresponds a counter-strategy of the opponent, having the same null-rank squares.

To active strategies and counter-strategies there also correspond specific moves. For instance, the open-file strategy requires moves of major pieces into the file, and the preparation for these moves—new attacks by major pieces on the file, and the freeing up of squares for them (if such a square is occupied by another major piece, that piece must stay within the file). The strategy of the center requires castling and its preparation—clearing pieces from the line joining the King and the Rook, the flanking moves of the Bishop (Bc1—b2, Bf1—g2 for White; Bc8—b7, Bf8—g7 for Black) and the preparation for these, namely b2—b3 for a Bishop on c1, or g2—g3 for a Bishop on f1, and so on. There are also specific moves for the counter-strategies. The counter to an attack on the King may be an arbitrary move by the King; if the opponent has major pieces, the counterstrategy may construct escape hatches, as it may against the open file strategy. Specific moves are defined by using the same predicates as were used to define general moves. For example, the moves specified above for the open file strategy are defined by the predicates

$$([L \cap G'']\&\urcorner[L \cap G'])\&[N \cap S_l],$$

$$\bigvee_{i \text{ a major piece}} ([L \cap M''[i]]\&\urcorner[L \cap M'[i]]),$$

$$\left[L \cap G' \cap \bigcup_{i \text{ a major piece}} M'_h[i]\right]\&\urcorner[N \cap S_l],$$

$$[L \cap G'']\&\left[L \cap G' \bigcup_{i \text{ a major piece}} M_h[i]\right]\&[N \cap S_l],$$

where the index i ranges over the major pieces, L is the board for the file under consideration, $S_l = (011000)$ is a six-bit array in which the one-digits occupy the places of major pieces—Queen and Rook. The $M_h[i]$ are the boards for the horizontal potentials of the major pieces.

Strategic moves, as defined above, are basically those that lead to exchanges on squares of rank 0 or 1. When bringing a short-range piece (King, Knight) in from a distance, to a key square of a strategy, we use the natural metric of the board: a piece approaches if the sum of the absolute values of the horizontal and vertical differences between the coordinates of the square on which the piece stands and the null-rank square decreases. A Pawn approaches the scene of action in accordance with the strategy of Pawn advance. For transferring long-range pieces (Queen, Rook, Bishop) we use like Botwinnik the concept of the trajectory (cf. [7—9]). All these transfers, however, are considered as separate strategies connected in some way with the basic strategy that motivates them.

Most of the moves allowed by the rules of chess correspond to one or another strategy. Therefore, to shorten the search, not all strategies need be admitted. We need rules for prohibiting strategies and rules for admitting those previously prohibited. Such rules may be based on a semantic analysis of the position under study and on the amount of time the program has already spent on moves in the current game (under time pressure new active strategies are wholly inadmissible). One of the aims in using a game model with strategic moves is to prevent planless play, in which the choice of a move depends on chance circumstances that influence the value of the evaluation function in the terminal positions of the model.

One might hope that the following set of rules would meet the requirements: At the base position in the model, all strategies are admissible. After a move A, B corresponding to one or more active strategies, all active strategies to which it does not correspond are suppressed in the B-subtree, but if the move corresponds to some admissible counter-strategy, no active strategy is suppressed. In the C-subtree of the model, after the opponent's reply B, C, counter-strategies are prohibited if they counter some active strategy of the opponent to which his reply does not correspond. If however the move B, C fits one of the opponent's admissible counter-strategies, our own counter-strategies are not suppressed.

It turns out, however, that almost all moves fit some or other counter-strategy. In fact, most of one's own side of the board consists of low-rank squares for counter-strategies, and new attacks on one's own pieces often result from one's own moves. Therefore the above mechanism for excluding moves is not strong enough. One might try to limit the number of counter-strategic moves in a branch, separately for White and Black, or to admit such moves only as replies to moves by the opponent that correspond to one of his active strategies. To be sure, what to do about counter-strategic moves at the base position is not obvious, but one might for example allow all of them. There is a more complex, 'semantic', solution. When the search reaches a position A for the first time, all moves corresponding to active admissible strategies are allowed, plus perhaps a few moves that are counter-strategic with respect to an admissible active strategy of the opponent, provided these are highly enough regarded by the best move routine. While searching the B-tree after a move (A, B) corresponding to various active strategies, one suppresses active strategies to which that move does not correspond. If, however, one is studying a counter-strategic move, no new prohibitions of active strategies occur in the search of the corresponding subtree.

Now suppose that we have returned to A after investigating the move (A, B). If the move turned out to be a refutation, nothing remains to be done, since we immediately step backward. If it is a bad move, we investigate the next of the moves permitted earlier. If, however, it is an improving move, that is, if it yields a partial score for A lying between the old bounds $\underline{\lim}$ and $\overline{\lim}$, we change the set of admissible counter-strategies. Initially the set is empty; the admissibility of a counter-strategic move in the array of best moves does not mean that other moves corresponding to the same counter-strategies are admissible. In the case we are now considering there is a critical branch from the position A beginning with the move (A, B) just investigated. From there onward, counter-strategies to active strategies of the opponent are admissible if they remain so at the end of the critical branch; if others were admissible earlier, they are now suppressed. However, if all the opponent's moves on the critical branch are counter-strategic, all our counter-strategies are suppressed.

If all the available moves have been investigated and have turned out to be bad, there exists a left W- or B-tree with base position at A, depending on whether White or Black has the move there. Then we also admit counter-strategies to any active strategies of the opponent that are admissible at the termini of the tree. However, if there are enough counter-strategic moves in the branch leading from the base position A_0 to the position A, and if we have found a satisfactory move at one of the lower rank positions in the branch, at which we have the move, no counter-strategies are admitted and we step backward from A.

The difficulty in carrying out such a plan is that at the precise moment when we admit some counter-strategy many moves (A, B) from A may have been studied and even rejected because they did not correspond to the strategies then admissible, or these moves may have been admitted but with a restriction on the set of admissible strategies. Therefore, after the set of admissible counter-strategies is changed, we must inspect these moves again. Those that correspond to newly admitted counter-strategies must be let into the search; some may have been judged admissible earlier, and if they were then held to be counter-strategic there is no need to inspect them again. However, they may have corresponded to some active strategies, and in this case such strategies must be prohibited and only those active strategies at A to which the moves in question do not correspond may be admitted. In any case, we cannot avoid repeated inspection of some variations in this model.

On the other hand, some of the strategies link together. Advancing a Pawn is useful for supporting neighboring Pawns (but not to outrun them); an attack on the King is supported by a corresponding Pawn advance; the strategy of the open file by the advance of the bordering Pawns, often necessary when the opponent has a Pawn on the file in question. Such a linkage can be realized by means of the graph Ind of induced strategies. If the arc $(S_1, S_2) \in \text{Ind}$, the admissibility of the strategy S_1 implies the admissibility of S_2. If S_2 is induced, the moves that answer to it must be counted as also answering to the inducing strategy S_1.

Transformations of strategies are also useful. For example, if the strategy of the open file leads to an invasion by major pieces, we should adopt a supplementary strategy of an attack on the King; or, after stationing our pieces on a null-rank square that is the next point on a trajectory, we should consolidate our elementary strategy and go over to the strategy of reaching the next point on the trajectory. The conditions for admissibility of a new strategy may also depend on changes in the position that are not directly connected with the aims of our strategies. The open file strategy may be admitted if a file opens up; a Pawn advance strategy when a passed Pawn eventuates, and so on. A change in the material balance may induce a review of formerly active or protective strategies, and the review may generate new admissions and supppressions in the subtrees rooted at the positions where the material balance changes.

Some of the above proposals have been implemented. A small practical experiment has shown that there are prospects for semantic models of the kind we have described, but fundamental research is still in the future.

CHAPTER 3

The Method of Analogy

Identical Moves in Different Positions

We have said that a game-playing algorithm often inspects the same thing many times over. A human, having studied a situation once, will in the future draw conclusions by the use of analogy. But, it often happens that seemingly insignificant changes in the position alter the course of the game and lead to substantially different outcomes. Such changes are said to be essential with respect to the contemplated variations. A human decides, well or poorly, whether a position that has been studied differs essentially from one that has not, and accordingly does or does not investigate variations starting from the latter. If we are to devise algorithms that use this method, we must analyze a) the notion of analogous moves (later we shall often use the term 'the same' rather than 'analogous') and b) the notion of the difference between positions essential for given variations.

We begin with the notion of similar moves. In the games that we have used and shall use as examples, pieces are moved about on a board and the moves are elementary displacements determined by a finite set of rules which is small compared to the number of positions, or even compared to the number of classes of mutually similar moves. To make the definition of a move more universal, we may consider an auxiliary square in addition to the customary squares of the board; in chess, for example, or draughts, captured pieces will move to the auxiliary square. The examples of moves that we have just given are taken from chess or draughts, but we may also imagine noughts-and-crosses as played on a board divided into squares; the pieces do not move on this board, but move to squares on it from an auxiliary square. Card games may be described in a similar way, e.g. bridge is played on six squares—N, S, E, W, the 'table', and an auxiliary square to which pieces move on quitting further play.

We may use a standard form for prescribing the rules of a game, by defining the admissible moves. Positions in the game are described by a set of incidence relations that hold among the elements of finite sets: T—the board (its elements are called squares); F—the set of pieces; and C—the set of players of the game. The relationship f/t is expressed in words as 'the piece f occupies the square t'; f/c as 'the piece f belongs to player c' (in many cases the relationships f/c do not change, i.e. each piece has its own 'color'). Each piece $f \in F$ in a given position stands on one square of the board and belongs to one of the players in the game. An arbitrary admissible move is the realization of a *virtual move*—a change in one of the incidence relationships. The lowest level in the hierarchy of rules describes the set of virtual moves.

A virtual move may be prescribed by specifying the following sets of incidence relationships:

(1) The set $\{f_i/t_i', f_i/c_i'\}$ which are satisfied before the move and not satisfied after it;
(2) the set $\{f_i/t''_i, f_i/c''_i\}$, which are satisfied after the move but not before it;
(3) the set $\{f_i/t'''_i, f_i/c'''_i\}$, which must be satisfied before the move (else the contemplated virtual move from the given position is inadmissible) and continue to be satisfied after the move.

Often the third set of relationships expresses the condition that some set of squares on the board must be empty if the virtual move is to be admissible. The squares c_i that satisfy the incidence relationships in the third set form a subset $L \subset T$ which we call the 'line of the move' or the 'trajectory'; (the line contains no squares when the third set is empty). In the first approximation virtual moves are admissible from positions at which the three given sets of incidence relationships are satisfied, the remaining incidence relationships being arbitrary. A real move is an admissible virtual move. One of the admissibility conditions relates to the turn to move. others are specified by using the notion of virtual move and some other notions found at higher levels in the hierarchy of rules.

Intuitively we would say that real moves are similar if they are realizations of the same virtual move. In specific games, however, we must take account of various circumstances belonging to higher levels, e.g. whether a check to a King arises or vanishes in consequence of a virtual move. Also, we may find it desirable to regard certain real moves as similar when the virtual moves they correspond to are different but have lines that are in some sense near together. In general, we shall hold to the intuitive definition.

Before we study the notion of essential differences between positions we pause to ask what kinds of situations make it desirable to reason by analogy in reaching a decision. Let B be the current position in a search. If (B, B') is a best move, we want to ascertain its consequences as precisely as we can. If it is a refutation, we must establish the fact that it is, in order to step backward from B without investigating other moves (B, C) (if we know that we may step back from B without studying any of the moves from it, then very probably we may omit a forward step to it). Moreover, since best and refutation moves can be relatively well guessed, the majority of the moves found in the search tree are bad. Therefore, to shorten the search, methods for mass pruning of bad moves are extremely valuable, and among these is the use of analogy.

To this end we consider the following scheme for reasoning by analogy. Suppose that at two positions B and C in which the move belongs to the same color, say White, we may make the similar moves (B, B_1) and (C, C_1). A search of the B_1-subtree of the game $\mathfrak{A}$ has shown that $\mathrm{sc}(B) \leq M$, and the corresponding search subtree $\tilde{\mathfrak{A}}(B)$ includes the move (B, B_1). The positions B, C and the tree $\tilde{\mathfrak{A}}(B)$ determine the value of the *influence* predicate $\mathrm{Inf}(B, C, \tilde{\mathfrak{A}}(B))$. The value 1 means that the difference between B and C influences the search tree $\tilde{\mathfrak{A}}(B)$. Then we must investigate the consequences of the move $C, C_1)$. The value 0 means that there is no influence, and the argument by analogy correctly concludes that $\mathrm{sc}(C_1) \leq m$.

To define the influence predicate so that its value can be determined without a search of the C_1-subtree, we examine the case $\mathrm{sc}(C_1) > m$. We assume for simplicity that White and Black move alternately. We also call positions with scores greater than m won positions, those with scores not greater than m lost, and regard similar moves as identical. Since a search

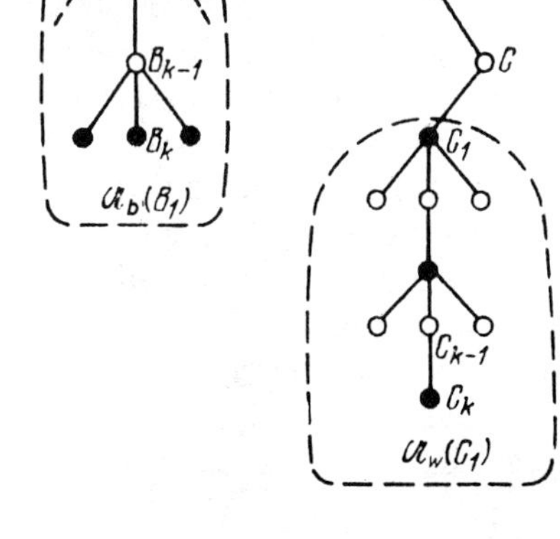

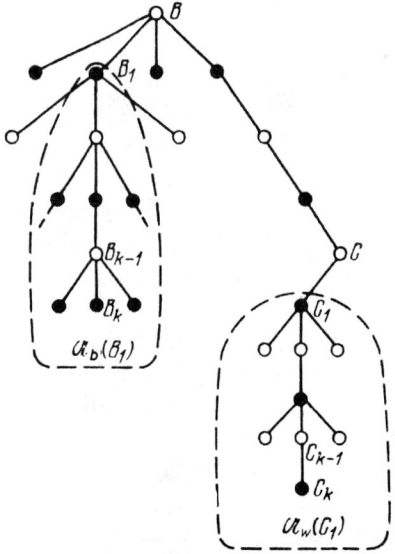

Figure 18

has shown that B_1 is lost, there exists a minimally inclusive B-pruned tree $\mathfrak{A}_b(B_1)$ in which all terminal positions are lost. So also there exists a W-pruned tree $\mathfrak{A}_w(C_1)$ with base C_1, in which the terminal positions are won, and in which there is at least one move from every non-terminal White position. To be sure, in contrast to the tree $\mathfrak{A}_b(B_1)$, this one is unknown, but for the present we do not need to know it.

The tree $\mathfrak{A}_b(B_1)$ contains a move $\Psi_1 = (B_1, B_2)$ which wins for Black and leads to a White position B_2. Suppose this same move is admissible at C_1 under the rules of the game $\mathfrak{A}$. It leads to some C_2, also a White position. Since $\mathfrak{A}_w(C_1)$ is a W-pruned tree and C_1 is a Black position this tree contains all moves in the game $\mathfrak{A}$ from C_1, including the move $\Psi = (C_1, C_2)$, and so contains the position C_2. If this is non-terminal the tree $\mathfrak{A}_w(C_1)$ contains a move $\Psi_2 = (C_2, C_3)$ leading from it to a Black position just as the move (B_2, B_3) does from the position $B_2 \in \mathfrak{A}_b(B_1)$ provided the latter move exists. As an arbitrary move from a White position in the tree $\mathfrak{A}_b(B_1)$, $(B_2, B_3) \in \mathfrak{A}_b(B_1)$ (see Figure 18).

We may repeat this argument to show that there exist branches $(B, B_1, \ldots, B_k)$ and $(C, C_1, \ldots, C_k)$ beginning at the positions B and C, respectively, and containing the same moves $\Psi_0, \Psi_1, \Psi_{k-1}$ in the same order (Ψ_0 is the move (B, B_1) which by hypothesis is identical to the move (C, C_1)), and the branch $(B, B_1, \ldots, B_k)$ belongs to the tree $\tilde{\mathfrak{A}}(B)$ and to its subtree $\mathfrak{A}_{\Psi_0}(B) = (B, B_1) \cup \mathfrak{A}_b(B_1)$. Since the trees $\mathfrak{A}_b(B_1)$ and $\mathfrak{A}_w(C_1)$ are finite, the process of constructing the sequence of moves $\Psi_0, \Psi_1, \ldots, \Psi_{k-1}$ must come to an end. The impossibility of continuing can arise only from one of the following causes:

(1) a Black move $J_k = (B_k, B_{k+1})$ which wins for Black at the position $B_k \in \mathfrak{A}_b(B_1)$, is impossible at the corresponding position;
(2) A White winning move $\Psi_k = (C_{k-1}, C_k) \in \mathfrak{A}_w(C_1)$ is impermissible at the position B_k;
(3) the position B_k is terminal and C_k is not;
(4) the position C_k is terminal and B_k is not;
(5) both B_k and C_k are terminal.

Our main interest lies in the first two cases, depicted in Figure 19. The last three cases may be excluded by a formal method. For games with two outcomes (which we are now studying since we regard all positions with scores not exceeding m as lost and positions with scores exceeding m as won) we can construct an equivalent model in which terminal positions are always lost for the side that has the move there. To do this we need only add to the game a fictitious move $(F, \bar{F})$ at all terminal positions F that are won for our own side; the new position $\bar{F}$ has the same score as F, i.e. is lost for the side having the move.

If the game $\mathfrak{A}$ satisfies our condition, C_k is a White position in Case 3 and there is a winning move $\Psi_{k+1} = (C_k, C_{k+1})$ from it. But B_k is terminal, and all moves, Ψ_{k+1} included, are prohibited from it. Thus in Case 3, Case

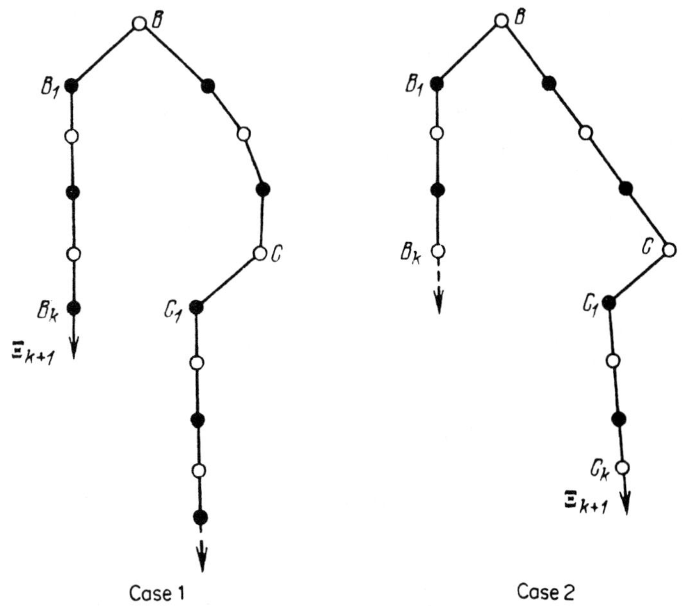

Figure 19

2 also holds; similarly, Case 1 holds when Case 4 does. Case 5 is impossible, since it implies different scores at the terminal positions B_k and C_k, where the same side has the move. This formal exclusion of Case 5, however, is useless for our investigation, since the fictitious move is defined quite differently from the genuine moves, and does not lend itself to the use of our notion of virtual move and our method of defining it.

But, it is especially important to exclude Case 5. Cases 3 and 4 may be studied by methods like those we shall use later for Cases 1 and 2. We shall admit some virtual moves from the positions B_k and C_k but not from other positions. In Case 5 we would have to use rules for determining the score at terminal positions; these would be complex, involving both real and virtual moves. Therefore we shall for the moment consider some more or less general examples of rules that result in the exclusion of Case 5 (and partly of Cases 3 and 4).

Suppose that a win or loss is determined by the configuration of only some of the pieces, and the pieces playing a role in the final configurations that determine a loss for White at terminal positions of $\mathfrak{A}_b(B_1)$ are identically placed in the positions B and C. Then the sequence of moves that leads from B_1 to the terminal position also leads from C_1 to a terminal position with the same score. For instance, in the game of 'Wolf and Sheep' the Wolf wins if he reaches the opponent's shelter in the first rank, just as in Russian draughts. Thus for this game both Cases 3 and 5 are impossible.

Furthermore, winning terminal configurations generally arise only from a move made by the winning side. In the game of noughts-and-crosses, for instance, such a configuration exists after the winner's move—five of his pieces placed sequentially in a vertical, horizontal, or diagonal line. We may modify the rules of this game to some extent, so that the move number becomes uniquely related to the score of the corresponding terminal positions, and Cases 3, 4, and 5 become impossible. It suffices that after the move number $L+1$ an arbitrary position is drawn if neither side has won earlier. In games such as draughts and chess a single-valued relationship between the turn to move and the score of the corresponding terminal position is prevented by the possibility of a draw based on a negative definition (neither side can force a win).

Another cause for the impossibility of Case 5 in many games, including of course model games, is that the score of their terminal positions is equal to the sum of the material gains achieved at each move in the sequence $\Psi_1, \Psi_2, \ldots, \Psi_k$ leading from the base position A_0 to the terminal position F:

$$\operatorname{sc}(F) = \sum_{i=1}^{k} h(\Psi_i).$$

For example, in many card games (and in *odnomastka*, a similar model game with complete information,) the scores are equal to the weighted sum of the tricks—sets of pieces simultaneously laid on the table. The weights of such tricks have opposite values for White and Black, and in some games depend on the pieces they contain.

In games such as draughts and chess the scores at terminal positions are immediately determined, in another way. Nevertheless, to achieve a winning position in these games it is often necessary to gain a preliminary material preponderance. Therefore, in computing the scores at terminal positions in models of these games, an important component is the material score discussed in the preceding chapter:

$$f_{\mathrm{m}}(C) := \sum_{\mu=1}^{M} h_\mu(C) P_\mu^w(C) - \sum_{\mu=1}^{M} h_\mu(C) P_\mu^b(C).$$

If the sequence of moves $\Psi_1, \Psi_2, \ldots, \Psi_k$ leads from the posiition B to the terminal position F, the material score of the latter is given by

$$\begin{aligned} f_M(F) &= f_M(B) + (f_M(F) - f_M(B)) \\ &= f_M(B) + \sum_{\mu=1}^{M} \big(P_\mu^b(F) - P_\mu^b(B)\big) h_\mu - \sum_{\mu=1}^{M} \big(P_\mu^w(F) - P_\mu^w(B)\big) h_\mu \\ &= f_M(B) + \sum_{i=0}^{\lceil k/2 \rceil} h(\Psi_{2^i+1}) - \sum_{i=1}^{\lceil k/2 \rceil} h(\Psi_{2^i}), \end{aligned}$$

where $h(\Psi)$ is the weight of the move Ψ. This is equal to the sum of the weights of the opponent's pieces that have been captured and the difference of the new and former weights of our own pieces that have changed position. This latter quantity differs from 0 for moves with promotion to King (draughts), Queen (chess), etc. For moves Ψ without capture or promotion, $h(\Psi)=0$.

Thus if the variation $(\Psi_0, \Psi_1, \ldots, \Psi_{k-l})$ leads from the position B to a material loss ($f_m(B_k) < f_m(B)$) it leads to the same loss from the position C. If the change in the positional score cannot compensate for White's material loss after the move $\Psi_0 = (B, B_1)$ from B, and $f_m(B) = f_m(C)$, then C_k, which corresponds to the terminal position $B_k \in \mathfrak{A}_{\Psi_0}(B)$, is necessarily lost for White. This means that Case 5 is impossible in our models. Cases 3 and 4 are excluded for these models because all terminal positions are lost for their own side, except those arising after the so-called *blank* moves. We shall discuss this point in more detail in the next section.

Thus, for a rather broad class of games the exclusion of Cases 1 and 2 at pairs of positions $B_k \in \mathfrak{A}_{\Psi_0}(B)$ andC_k belonging to the 'parallel' tree $\mathfrak{A}_{\Psi_0}(C)$ implies that $\mathrm{Inf}(B, C, \mathfrak{A}_{\Psi_0}(B)) = 0$. Therefore we may construct the influence predicate by studying the conditions that allow the difference in the simultaneous admissibility of the same move in parallel positions. Let $\mathrm{Inf}(B, C, D)$ be a predicate having the value 1 if a) Black has the move at $D \in \mathfrak{A}_{\Psi_0}(B)$ and the move $\Psi = (D, G) \in \mathfrak{A}_{\Psi_0}(B)\}$ *is inadmissible at the parallel position* $E \in \mathfrak{A}_{\Psi_0}(C)$, or b) White has the move at D and there exists a virtual move $\Psi(E, H)$ which is admissible at the parallel position $E \in \mathfrak{A}_{\Psi_0}(C)$ and inadmissible at D. The influence predicate $\mathrm{Inf}(B, C, \mathfrak{A}_{\Psi_0}(B))$ is easily expressed by means of such a 'local' predicate:

$$\mathrm{Inf}\big(B, C, \mathfrak{A}_{\Psi_0}(B)\big) = \vee\big\{\mathrm{Inf}(B, C, D) \mid D \in \mathfrak{A}_{\Psi_0}(B)\big\}.$$

Since an arbitrary position is fully defined by a finite collection of features, i.e. by the values of elementary predicates belonging to a finite set $\{\pi_\alpha\}$, the predicate $\mathrm{Inf}(B, C, D)$ can be represented in disjunctive normal form with components $\pi_\alpha(B)$, $\pi_\alpha(C)$, and $\pi_\alpha(D)$. (We may assume that the set $\{\pi_\alpha\}$ contains the negation of each of its predicates.) This form can be written in terms of the non-elementary predicates of the positions B, C, and D:

$$\mathrm{Inf}(B, C, D) = \bigvee_{i=1}^{s} \big(\rho_i(B) \& \sigma_i(C) \& \tau_i(D)\big).$$

Consequently

$$\begin{aligned}\mathrm{Inf}\big(B, C, \mathfrak{A}_{\Psi_0}(B)\big) &= \bigvee_{D \in \mathfrak{A}_{\Psi_0}(B)} \left(\bigvee_{i=1}^{s} \big(\rho_i(B) \& \sigma_i(C) \& \tau_i(D)\big)\right) \\ &= \bigvee_{i=1}^{s} \big(\rho_i(B) \& \sigma_i(C)\big) \& \left(\bigvee_{D \in \mathfrak{A}_{\Psi_0}(B)} \tau(D)\right) \\ &= \bigvee_{i=1}^{s} \Big(\rho_i(B) \& \sigma_i(C) \& \tau\big(\mathfrak{A}_{\Psi_0}(B)\big)\Big).\end{aligned}$$

for an arbitrary set of positions $\mathfrak{W}$

$$\tau_i(\mathfrak{W}) = \vee\{\tan_i(D) \mid D \in \mathfrak{W}\}, \qquad i=1,2,\dots,s.$$

The predicates $\tau_i(\mathfrak{W})$ have an important property:

$$\tau_i(\mathfrak{W}_1 \cup \mathfrak{W}\,2) = \tau_i(\mathfrak{W}1) \vee \tau_i(\mathfrak{W}_2).$$

Thus the predicates $\rho_i(B)$ and $\sigma_i(C)$ may be determined for the positions B and C, while the values of the predicates $\tau_i(\mathfrak{A}_{\Psi_0}(B))$ can be calculated during the search of the tree $\mathfrak{A}_{\Psi_0}(B)$ and the amount of information about them does not depend on the number of positions in the tree. If the predicate $\mathrm{Inf}(B, C, D)$ is *exact*, i.e. if it is a necessary and sufficient condition for the occurrence of Cases 1 and 2, the number s of predicates may be very large, so that the computation of the predicates $\tau_i(\mathfrak{A}_{\Psi_0}(\mathrm{B}))$ consumes large amounts of time and memory. However, as we shall see, we can develop predicates for specific games that can be computed quickly enough, if we replace the necessary and sufficient conditions for the existence of Cases 1 and 2 by necessary conditions only.

Suppose that a search of the tree $\mathfrak{A}_{\Psi_0}(B)$ has shown that the move $\Psi_0 = (B, B_1)$ leads to a lost position, and after the moves $\Theta_1, \Theta_2, \dots, \Theta_{2l}$ we arrive from B at a position S which is also White. Fairly often some new moves $\Omega_1, \Omega_2, \dots, \Omega_r$ can be made from C, although they were inadmissible at B. Then the influence predicate $\mathrm{Inf}(B, C, \mathfrak{A}_{\Psi_0}(B))$ is equal to 1, since the position C in the tree $\mathfrak{A}_{\Psi_0}(C)$ corresponds to the position B in the tree $\mathfrak{A}_{\Psi_0}(B)$ and we are in the presence of Case 2. Replacing the trees $\mathfrak{A}_{\Psi_0}(B)$ and $\mathfrak{A}_{\Psi_0}(C)$ by $\mathfrak{A}_{\Psi_0}(B_1)$ and its parallel tree $\mathfrak{A}_{b\parallel}(C_1)$ normally does not help us, since some at least of the new moves Ω_λ remain admissible at White positions in the tree $\mathfrak{A}_{b\parallel}(C_1)$. Thus if we use the notion of influence as defined above we cannot shorten the search of the C-subtree of $\mathfrak{A}$.

Also, the new moves Ω_λ $(1 \le \lambda \le r)$ may be irrelevant to the play and will therefore lose quickly. Common sense tells us that in this case our earlier conclusion about the move Θ_0, which is admissible at C also, will probably still be valid. We need to reconsider it only if the sequence of moves $\Lambda(\Theta_1, \Theta_2, \dots, \Theta_{2l})$—or variants of it—which demonstrate that White loses after the moves $\Omega_\lambda (1 \le \lambda \le r)$ turn out to influence variants of the tree $\mathfrak{A}_{\Psi_0}(B)$. But this is a different concept of influence. It is narrower, since now it prohibits not all instances of Case 2, but only those in which White moves that have not yet been inspected are admissible . The new influence predicate $\mathrm{Inf}'(B, C, \mathfrak{A}_b(B))$ is determined by the conditions that permit new moves, not in Ω_λ, to arise at positions in the tree $\mathfrak{A}_{\Psi_0}(C)$.

A scheme for shortening the search with this notion of influence is shown in Figure 20. First we investigate the new moves $\Omega_1, \Omega_2, \dots, \Omega_r$. Let $\mathfrak{A}_{\{\Omega_\lambda\}}(C)$ be the corresponding search tree, and let $\mathfrak{A}_{\{\Omega_\lambda\}}(B) = \Lambda \cup \mathfrak{A}_{\{\Omega_\lambda\}}(C)$. If the tree $\mathfrak{A}_{\{\Omega_\lambda\}}(C)$ has no influence on the tree $\mathfrak{A}_{\Psi_0}(B)$, the move $\Psi_0 = (B, B_1) = (C, C_1)$ from the position C loses just as it does from the position B. (A more precise description of this method for shortening the search will be

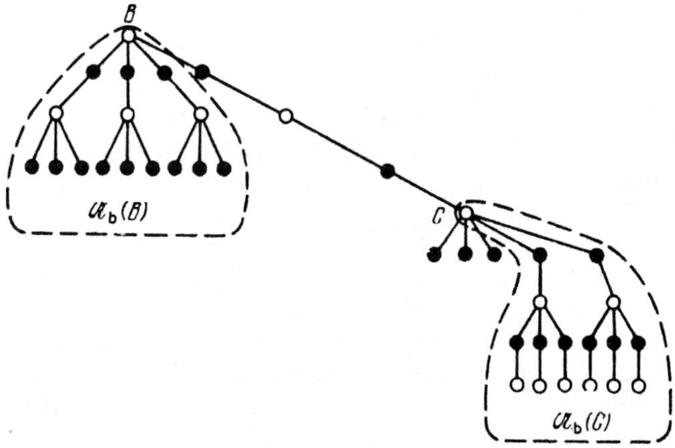

Figure 20

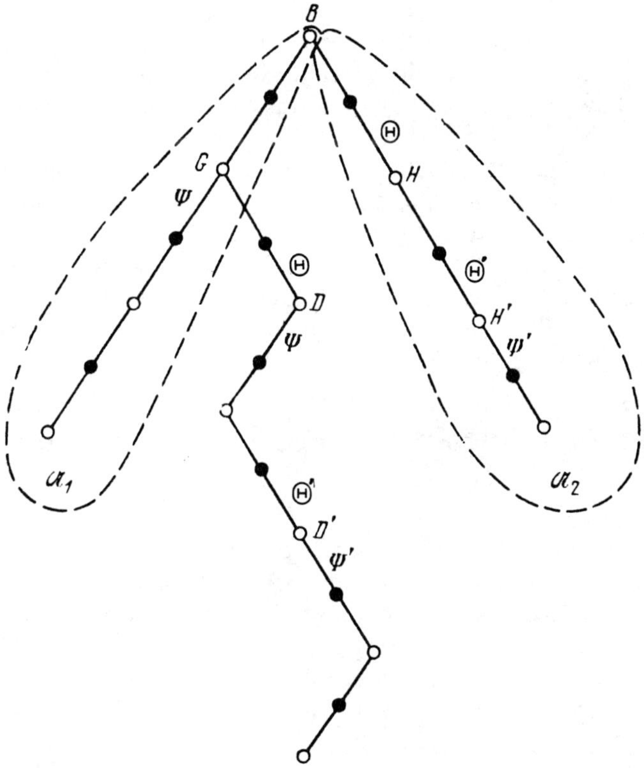

Figure 21

given in Section 4 of this chapter.) Thus we need to determine the influence of a subtree $\mathfrak{A}_1$ with base B on another subtree $\mathfrak{A}_2$ with the same base, rather than the influence of two different positions B and C on the search tree $\mathfrak{A}_{\Psi_0}(B)$. An influence predicate $\mathrm{Inf}(B, \mathfrak{A}_1, \mathfrak{A}_2)$ of this kind is the disjunction of the influence predicates of the pairs of branches Λ_1 and Λ_2 in these subtrees:

$$\mathrm{Inf}(B, \mathfrak{A}_1, \mathfrak{A}_2) = \bigvee \{\mathrm{Inf}(B, \Lambda_1, \Lambda_2) \mid gL_1 \in \mathfrak{A}_1, \Lambda_2 \in \mathfrak{A}_2\}.$$

The B-subtree of the game $\mathfrak{A}$ contains compound branches, in which the moves of some branch $\Lambda_1 \subset \mathfrak{A}_1$ alternate with the moves of some other branch $\Lambda_2 \subset \mathfrak{A}_2$, while the relative order of the moves in the two branches is preserved (cf. Figure 21). Only one of the branches Λ_1, Λ_2 may have an odd length, else the numbers of White and Black moves in the compound branch Λ would differ by 2, which is impossible in a game where White and Black move alternately. The positions D in Λ that are incident on the moves $\Psi \in \Lambda_1$ and $\Theta \in \Lambda_2$ correspond to the positions $G \in \mathfrak{A}_1$ and $H \in \mathfrak{A}_2$ that are incident on the moves Ψ and Θ in the respective branches Λ_1 and Λ_2. When both the incident positions D' of a move belong to the same branch, Λ_1 or Λ_2, the position H' corresponding to D' lies in only one of the trees $\mathfrak{A}_1$, $\mathfrak{A}_2$. The final position in the branch Λ corresponds to the final position of the same color in either Λ_1 or Λ_2 when the lengths of these branches have unequal parities, or to the final positions in both branches when their lengths have even parity.

In this more general situation, Case 1 indicates that at some position D lying in the branch Λ, with color different from that of B, some move Ψ is inadmissible under the rules of the game $\mathfrak{A}$ but is admissible at the corresponding position G in one or other of the branches Λ_1 or Λ_2 and belonging to that branch; Case 2 indicates that some move Ω is admissible at the position D (of the same color as B) and inadmissible at both of the corresponding positions.

Axioms of Influence and Possibilities for Shortening the Search

We suppose now that the influence predicate $\mathrm{Inf}(B, \Lambda_1, \Lambda_2)$ has been defined as a function of the position B and two issuing branches Λ_1 and Λ_2. Its properties, as used to prove the theorem on the possibility of shortening the search, may be stated axiomatically, but before we state them we need some preliminary definitions.

The branch $\Lambda(\Psi_1, \Psi_2, \ldots, {}_{\Lambda})$ is said to be *composed of the two branches* $\Lambda_1(\Theta_1, \Theta_2, \ldots, \Theta_k)$ and $\lambda_2(\Omega_1, \Omega_2, \ldots, \Omega_{1-k})$:

$$\Lambda = \Lambda_1 + \Lambda_2,$$

if every move $\Psi_h \in \Lambda$ coincides with one of the moves $\Theta \in \Lambda_1$ or $\Omega \in \Lambda_2$, while the correspondence between the moves in the branch Λ and the moves in Λ_1 and Λ_2 is one-to-one, and the ordering of the moves in these branches is compatible, i.e.

$$\Theta_i = \Psi_h \& \Theta_{i+1} = \Psi_g \Rightarrow h < g, \qquad i = 1, 2, \ldots, k-1,$$

$$\Omega_j = \Psi_m \& \Omega_{j+1} = \Psi_n \Rightarrow m < n, \qquad j = 1, 2, \ldots, l-k-1.$$

The branch $\Lambda(\Psi_1, \Psi_2, \ldots, \Psi_1)$ is said to be *strictly composed of the two branches* $\Lambda_1(\Theta_1, \Theta_2, \ldots, \Theta_k)$and $\Lambda_2(\Omega_1, \Omega_2, \ldots, \Omega_{l-k})$:

$$\Lambda = \Lambda_1 * \Lambda_2,$$

if

$$\Psi = \begin{cases} \Theta_i, & i = 1, 2, \ldots, k, \\ \Omega_{i-k}, & i = k+1, k+2, \ldots, 1. \end{cases}$$

The non-commutative operation of strict composition, as so defined, may at times fail to define a branch in $\mathfrak{A}$ with its origin at B, since some move Ψ_i may be inadmissible under the rules of $\mathfrak{A}$ at the position arising after the sequence of moves $\Psi_1, \Psi_2, \ldots, \Psi_{i-1}$ (in this case clearly $i > k$). Nevertheless, the branch $\Lambda = \Lambda_1 * \Lambda_2$ may be admissible even when the branch Λ_2 issuing from B consists of moves not necessarily admissible at the corresponding positions. The non-strict composition $\Lambda = \Lambda_1 + \Lambda_2$ is not uniquely defined (the moves in the branches may alternate in various ways) and may also define an inadmissible branch. It may be regarded as commutative, in the sense that the relationship $\Lambda = \Lambda_1 + \Lambda_2$ among the branches is commutative in the second and third arguments. The strict composition of the branch Λ and a move Θ is defined as for branches:

$$\Lambda(\Psi_1, \Psi_2, \ldots, \Psi_l) * \Theta := \Lambda'(\Psi_1, J_2, \ldots, \Psi_l, \Theta).$$

Let B be a fixed position which will serve as the origin of all branches to be considered. If $\mathrm{Inf}(B, \Lambda_1, \Lambda_2) = 1$, we shall say that the branch Λ_1 influences the branch Λ_2 and we write $\Lambda_1 \sim \Lambda_2$. We shall also consider the influence relation $\Lambda \sim \Psi$ of the branch Λ on the move Ψ; this is not the same as the influence $\Lambda \sim L(\Psi)$ of Λ on the branch $L(\Psi)$ which consists of the single move Ψ. If the move Ψ is admissible at B, the relation $\Lambda \sim \Psi$ implies that $\Lambda \sim L(\Psi)$, but the converse is not true.

The position D to which the branch $\Lambda(\Psi_1, \Psi_2, \ldots, \Psi_l)$ leads from B will be denoted by $\mathrm{fin}(B, \Lambda)$. The color of the move Ψ will be denoted by $\mathrm{col}\,\Psi$; the color of the side having the move at A will be denoted by $\mathrm{col}\,A$, and the color of the side having the move at the position $\mathrm{fin}(B, \Lambda)$ by $\mathrm{col}\,\Lambda$. (Since we are currently considering games $\mathfrak{A}$ in which White and Black move alternately, $\mathrm{col}\,\Lambda$ is uniquely defined by $\mathrm{col}\,B$ and the parity of the length of the branch Λ.) When B is a White position we write $\mu \underset{\mathrm{col}\,B}{\preccurlyeq} \nu$ to mean that $\mu \leq \nu$, and $\mu \underset{\mathrm{col}\,B}{\prec} \nu$ to mean that $\mu + \delta \leq \nu$, where δ is some preselected quantity; when $\mathrm{col}\,B$ is Black, we use the same notation to mean that $\mu \geq \nu$ and $\mu - \delta \geq \nu$, respectively. If $\text{sc}(D) \underset{\text{col } B}{\prec} \text{sc}(E)$ we shall say that the score at the position D is significantly worse for col(B) than the score at E.

The set of virtual moves admitted by the rules of $\mathfrak{A}$ at some position A will be denoted by $M(A)$.

The Influence Axioms. In the following formulations we shall suppose that in the game $\mathfrak{A}$ White and Black move alternately, and that all branches issue from a fixed (arbitrary) position B and consist of moves admissible in $\mathfrak{A}$ at the corresponding positions.

Axiom 1. Axiom on the relationship between influence on a move and influence on a branch: If the branch Λ_1 influences the move Ψ with the color col B and Ψ belongs to the branch Λ_2, then Λ_1 influences Λ_2:

$$\Lambda_1 \sim \Psi \& \text{col } \Psi = \text{col } N \& \Psi \in \Lambda_2 \Rightarrow \Lambda_1 \sim \Lambda_2.$$

Axiom 2. Axiom of symmetry: If the branch Λ_1 influences the branch Λ_2, then Λ_2 influences Λ_1:

$$\Lambda_1 \sim \Lambda_2 \Rightarrow \Lambda_2 \sim \Lambda_1.$$

Axiom 3. First axiom on the composition of branches: If the branch Λ is composed of the branches Λ_1, Λ_2 and influences the move Ψ, then either Λ_1 influences Ψ or Λ_2 influences Ψ or Λ_1 influences Λ_2.

$$\Lambda = \Lambda_1 + \Lambda_2 \& \Lambda \sim \Psi \Rightarrow \Lambda_1 \sim \Psi \vee \Lambda_2 \sim \Psi \vee \Lambda_1 \sim \Lambda_2.$$

Axiom 4. The second axiom on the composition of branches: If the branch Λ is composed of the branches Λ_1 and Λ_2 and influences the branch Λ_3, then Λ_1 influences Λ_3 or Λ_2 influences Λ_3 or Λ_1 influences Λ_2:

$$\Lambda = \Lambda_1 + \Lambda_2 \& \Lambda \sim \Lambda_3 \Rightarrow \Lambda_1 \sim \Lambda_3 \vee \Lambda_2 \sim \Lambda_3 \vee \Lambda_1 \sim \Lambda_2.$$

Axiom 5. Axiom on Case 1: Let Λ_1, Λ_2 be branches with even and odd lengths, respectively. Let $\Lambda = \Lambda_1 + \Lambda_2$, and let the move Ψ be admissible under the rules of the game $\mathfrak{A}$ at the position fin(B, Λ_2) and inadmissible at fin(B, Λ) (see Figure 22). Then the branch Λ_1 influences $\Lambda_2 * \Psi$:

$$\begin{aligned} \text{col } \Lambda_1 = \text{col } B \& \text{col } \Lambda_2 \neq \text{col } B \& \Lambda \\ = \Lambda_1 + \Lambda_2 \& \Psi \in M(\text{fin}(B, \Lambda_2)) \& \\ \Psi \notin M(\text{fin}(B, \Lambda)) \Rightarrow \Lambda_1 \sim \Lambda_2 * \Psi. \end{aligned}$$

Axiom 6. First axiom on Case 2. Let Λ_1 and Λ_2 be two branches of even length, let $\Lambda = \Lambda_1 + \Lambda_2$, and let the move Ψ be inadmissible under the rules of the game $\mathfrak{A}$ at the positions fin(B, Λ_1) and fin(B, Λ_2) but admissible at fin(B, Λ) (see Figure 23). Then Λ_1 influences Λ_2:

$$\begin{aligned} \text{col } \Lambda_1 = \text{col } \Lambda_2 = \text{col } B \& \Lambda = \Lambda_1 + \Lambda_2 \\ \& \Psi \notin (M(\text{fin}(B, \Lambda_1)) \cup M(\text{fin}(B, \Lambda_2)) \\ \& \Psi \in M(\text{fin}(B, \Lambda)) \Rightarrow \Lambda_1 \sim \Lambda_2. \end{aligned}$$

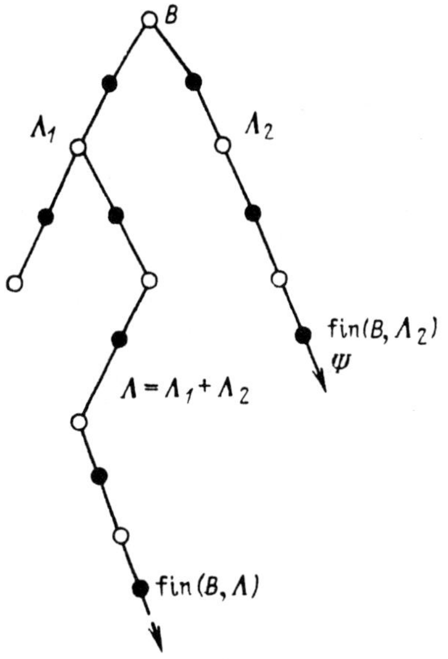

Figure 22

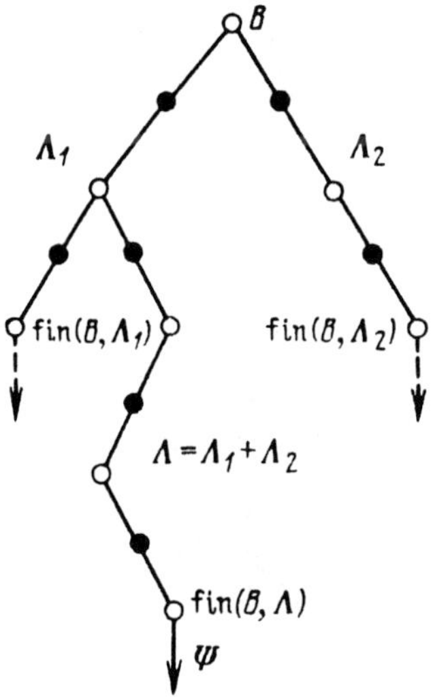

Figure 23

Axiom 7. Second axiom on Case 2. Let Λ_1 and Λ_2 be branches of even length, let $\Lambda = \Lambda_1 + \Lambda_2$, and let the move Ψ be admissible under the rules of the game A at the positions B, $\operatorname{fin}(B, \Lambda_1)$, and $\operatorname{fin}(B, \Lambda)$, but not admissible at the position $\operatorname{fin}(B, \Lambda_2)$ (see Figure 24). Then Λ_1 influences Λ_2 or Ψ:

$$\begin{aligned} &\operatorname{col} \Lambda_1 = \operatorname{col} \Lambda_2 \\ &\qquad = \operatorname{col} B \& \Lambda = \Lambda_1 + \Lambda_2 \& \Psi \in \big(M(B) \cap M(\operatorname{fin}(B, \Lambda_1)) \cap M(\operatorname{fin}(B, \Lambda))\big) \\ &\quad \& \Psi \notin M(\operatorname{fin}(B, \Lambda_2)) \Rightarrow \Lambda_1 \sim \Lambda_2 \vee \Lambda_1 \sim \Psi. \end{aligned}$$

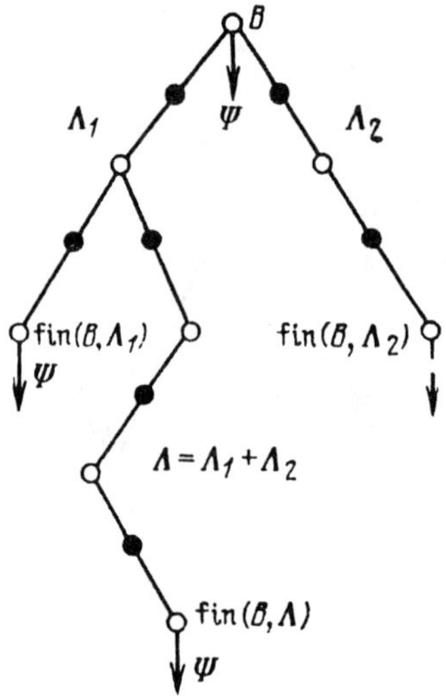

Figure 24

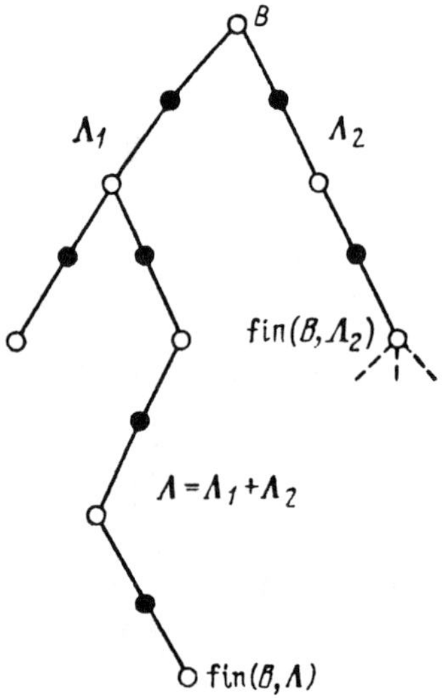

Figure 25

Axiom 8. Axiom on Cases 3 and 4. Let Λ_1 be a branch of even length, let $\Lambda = \Lambda_1 + \Lambda_2$, and let one and only one of the positions $\text{fin}(B, \Lambda)$ and $\text{fin}(B, \Lambda_2)$ be terminal (see Figure 25). Then Λ_1 influences Λ_2:

$$\begin{aligned}\text{col}\,\Lambda_1 = \text{col}\,B \& \Lambda = \Lambda_1 \dotplus \Lambda_2 \\ \& M(\text{fin}(B, \Lambda)) = \varnothing \& M(\text{fin}(B, \Lambda_2)) \neq \varnothing \\ \vee M(\text{fin}(B, \Lambda)) \neq \varnothing \& M(\text{fin}(B, \Lambda_2)) = \varnothing \Rightarrow \Lambda_1 \sim \Lambda_2.\end{aligned}$$

Axiom 9. Axiom on Case 5. Let Λ_1 be a branch of even length, let $\Lambda = \Lambda_1 + \Lambda_2$, and let both positions $\text{fin}(B, \Lambda)$ and $\text{fin}(B, \Lambda_1)$ be terminal

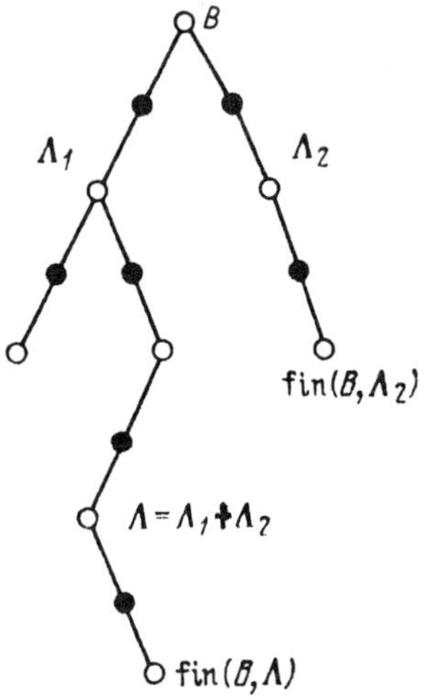

Figure 26

(see Figure 26). Then either Λ_1 influences Λ_2 or the score at the position $\operatorname{fin}(B, \Lambda)$ is not better for col B than the score at $\operatorname{fin}(B, \Lambda_2)$:

$$\begin{aligned}\operatorname{col}\Lambda_1 &= \operatorname{col} B \& \Lambda = \Lambda_1 + \Lambda_2 \& M(\operatorname{fin}(B, \Lambda)) \\ &= M(\operatorname{fin}(B, \Lambda_2)) = \varnothing \\ &\Rightarrow \Lambda_1 \sim \Lambda_2 \vee \operatorname{sc}(\operatorname{fin}(B, \Lambda)) \underset{\operatorname{col} B}{\preccurlyeq} \operatorname{sc}(\operatorname{fin}(B, \Lambda_2)).\end{aligned}$$

Constructively defined influence relations for specific games satisfy these axioms so to speak 'essentially', that is, when their assertions are invalid then the contemplated branches, their ends, or the moves will satisfy supplementary conditions that we can specify or exploit. Thus the formulations of the theorems we shall be proving must be sharpened, and their proofs must be supplemented by investigations of special cases. Nevertheless the leading ideas of our proofs will remain the same as those in this chapter. Of course, this system of concepts and axioms connected with the influence relation is not in final form.

In particular, for draughts, chess, similar games, and many models of similar games, we may replace the axioms for Cases 3, 4, and 5 by conditions that hold when these cases arise. The conditions are a consequence of certain properties of such games that we shall now discuss.

(1) For every position A in the contemplated games there is a function yielding the material balance $f_{\mathrm{m}}(A)$.
(2) For any position $B \in \mathfrak{A}$ and for any branch Λ issuing from it that consists of moves admissible under the rules of $\mathfrak{A}$ at the corresponding positions, the condition

$$f_{\mathrm{m}}(\operatorname{fin}(B, \Lambda)) = f_{\mathrm{m}}(B) + \sum_{\Psi \in \Lambda} h(\Psi)$$

holds, where the weights $h(\Psi)$ are defined for all elements Ψ of the set V of virtual moves and do not depend on the positions at which these moves are made.

(3) The set V may contain blank moves of both White and Black, which we shall denote indifferently by the symbol $m_{\varnothing}$, since this leads to no confusion. The weight $h(m_{\varnothing})$ of a blank move is equal to 0, and a blank move always leads to a terminal position.

(4) If a terminal position $F \in \mathfrak{A}$ is the end of a non-blank move $(D, F) \neq m_{\varnothing}$, it either has a score which we may take as losing for col F (we recall that White and Black move alternately in $\mathfrak{A}$ and therefore the concept of the color of a terminal position makes sense), or it is a position of a special type that may in some approximations be ignored. For instance, White or Black wins at draughts when the opponent is either bereft of pieces or has all his pieces 'locked', unable to make a legal move. These cases arise when the opponent has the move. In exactly the same way White or Black is mated in chess, i.e. loses, in positions where the loser has the move.

The existence of a third outcome of the game—namely a draw—prevents a complete satisfaction of Condition 4; however, we may take as terminal positions with score $1/2$ only those in which the side that was seeking a win has the move. We may always do this in draughts, where a draw occurs when it can be proved that neither side can win. The same is true of chess, with a reservation excluding stalemated positions, either ignoring them completely or subjecting them to specific investigation (such investigations will enable us to prove even for chess the theorems we shall be stating later).

If there are no blank virtual moves in the game $\mathfrak{A}$, Property 4 suffices to exclude Cases 3, 4, and 5.

(5) If the blank move $(D, F) = m_{\varnothing}$ leads to the terminal position F, the score there is equal to the material balance:

$$(D, F) = m_{\varnothing} \Rightarrow \mathrm{sc}(F) = f_{\mathrm{m}}(F).$$

Let us see how this property is used in the investigation of Cases 3, 4, and 5. (To prove the theorem on the carry-over of the results of an estimate of a group of moves we need more complex arguments, but these are essentially close to those we shall adduce below.)

Case 3 means that $\Lambda = \Lambda_1 + \Lambda_2$, the position $\mathrm{fin}(B, \Lambda_1)$ is terminal and $\mathrm{fin}(B, \Lambda)$ is not. Accordingly neither of the branches Λ and Λ_1 contains a blank move, since by Property 3 a blank move leads to a terminal position. In such a situation, certain conditions on Case 2 are satisfied if Λ_1 has even length, and Case 2 may be investigated with the aid of these axioms. If Λ_1 has odd length Property 4 implies that the terminal position $\mathrm{fin}(B, \Lambda_1)$ is lost for col $\Lambda_1 \neq$ col B. In the situations we shall be studying this turns out to be impossible.

In Case 4 $\Lambda = \Lambda_1 + \Lambda_2$, the next move Θ is the same in both branches Λ and Λ_1, and the position $\mathrm{fin}(B, \Lambda)$ is terminal, while $\mathrm{fin}(B, \Lambda_1)$ is not. Then the move Θ, which leads to $\mathrm{fin}(B, \Lambda_1)$ is non-blank. When the branch Λ_1 has an odd length the conditions of the axiom on Case 1 are satisfied. But if this length is odd, Property 4 implies that the position $\mathrm{fin}(B, \Lambda)$ is lost for col Λ = col B,which is impossible in the cases that we considered in proving the theorem on the transfer of scores.

Finally, suppose that $\Lambda = \Lambda_1 + \Lambda_2$, that the next moves in the branches Λ and Λ_1 coincide, and that the positions $\mathrm{fin}(B, \Lambda)$ and $\mathrm{fin}(B, \Lambda_1)$ are both terminal; i.e. we have Case 5. If the next move Θ in the branches Λ and Λ_1 is non-blank, Property 4 implies that the positions $\mathrm{fin}(B, \Lambda)$ and $\mathrm{fin}(B, \ldots, \Lambda_1)$ are lost for the same side. If, however, $\Theta = m_{\varnothing}$, Property 5 implies that

$$\begin{aligned}\mathrm{sc}(\mathrm{fin}(B, \Lambda)) &= f_{\mathrm{m}}(\mathrm{fin}(B, \Lambda)) \\ &= f_{\mathrm{m}}(B) + \sum_{\Psi \in \Lambda} h(\Psi) \\ &= f_{\mathrm{m}}(B) + \sum_{\Psi \in \Lambda_1} h(\Psi) + \sum_{\Psi \in \Lambda_2} h(\Psi) \\ &= f_{\mathrm{m}}(\mathrm{fin}(B, \Lambda_1)) + \sum_{\Psi \in \Lambda_2} h(\Psi) \\ &= \mathrm{sc}(\mathrm{fin}(B, \Lambda_1)) + \sum_{\Psi \in \Lambda_2} h(\Psi).\end{aligned}$$

So, if

$$\sum_{\Psi \in \Lambda_2} h(\Psi) \underset{\mathrm{col}\,B}{\preccurlyeq} 0 \quad \text{and } \mathrm{sc}(\mathrm{fin}(B, \Lambda_1)) \underset{\mathrm{col}\,B}{\prec} m,$$

we have

$$\mathrm{sc}(\mathrm{fin}(B, \Lambda)) \underset{\mathrm{col}\,B}{\prec} m.$$

Thus we cannot have the position $\mathrm{fin}(B, \Lambda_1)$ lost for the side $\mathrm{col}(B)$ and $\mathrm{fin}(B, \Lambda)$ not lost for that side; however we must impose the supplementary constraint

$$\sum_{\Psi \in \Lambda_2} h(\Psi) \preccurlyeq 0.$$

We now introduce yet more notation. Let $\mathfrak{B}$ be a subtree of the game $\mathfrak{A}$, B its base, and $L(\Psi_1, \Psi_2, \ldots, \Psi_l)$ a sequence of virtual moves. We define the relation $L \diamond B$; the sequence L defines a branch with base B consisting of moves admissible under the rules of $\mathfrak{A}$ at the corresponding positions and belonging to the subtree B, i.e.

$$\begin{aligned}L \diamond B := \Psi_1 = (B, B_1) \in \mathfrak{B}\,\& \\ \Psi_2 = (B_1, B_2) \in \mathfrak{B}\,\& \ldots \& \\ \Psi_l = (B_{l-1}, B_l) \in \mathfrak{B}.\end{aligned}$$

Then we say that the branch L is incident on the subtree $\mathfrak{B}$. Using the notion of incidence we can define the influence of a branch L on a subtree $\mathfrak{B}$ of the game tree $\mathfrak{A}$, and the influence of a subtree $\mathfrak{B}_1$ on a subtree $\mathfrak{B}_2$ (these subtrees must have a common root):

$$L \sim \mathfrak{B} := \exists \Lambda (\Lambda \diamondsuit \mathfrak{B} \& L \sim \Lambda);$$

$$\mathfrak{B}_1 \sim \mathfrak{B}_2 := \exists \Lambda_1 \exists \Lambda_2 (\Lambda_1 \diamondsuit \mathfrak{B}_1 \& \Lambda_2 \diamondsuit \mathfrak{B}_2 \& \Lambda_1 \sim \Lambda_2).$$

We shall consider a certain subtree $\mathfrak{S}(B)$ of the tree for our game $\mathfrak{A}$, and the corresponding model games. The position $B$is the root of the subtree and is the base position of the model game. For every position $D \in \mathfrak{S}(B)$ we denote by $S(D)$ the set of virtual moves to which the arcs $(D, E) \in \mathfrak{S}(B)$ correspond. Clearly $S(D) \subset M(D)$. The side $\mathrm{col}_{\mathfrak{S}(B)}(D)$ having the move at the position D in the game $\mathfrak{S}(B)$ and the scores at the positions $F \in \mathfrak{S}(B)$ that are terminal in the original game and therefore terminal in $\mathfrak{S}(B)$ are defined in the natural way:

$$\mathrm{col}_{\mathfrak{S}(B)}(D) := \mathrm{col}(D);$$

$$\mathrm{sc}_{\mathfrak{S}(B)}(F) := \mathrm{sc}(F).$$

If, however, F is terminal in $\mathfrak{S}(B)$ only, and not in $\mathfrak{A}$, its score in the model game is the least of those possible in $\mathfrak{A}$ when F is a White position and the greatest when F is a Black position.

$$F \in \mathfrak{S}(B) \& M(F) \neq \varnothing \& S(F) = \varnothing$$
$$\Rightarrow \forall D \subset \mathfrak{A} \left\{ \mathrm{sc}_{\mathfrak{S}(B)}(F) \underset{\mathrm{col}\, B}{\preccurlyeq} \mathrm{sc}(D) \right\}.$$

Let $\mathfrak{S}(C)$ be such a model game. In particular, we shall allow all its sets $S(D)$ to coincide with the set $M(D)$ of virtual moves admissible in $\mathfrak{A}$ at the corresponding positions. Then $\mathfrak{S}(C)$ is defined by a C-subtree of $\mathfrak{A}$, which we shall denote by $\mathfrak{A}(C)$. The model game $\mathfrak{S}_{\mathrm{T}}(B)$ is called a *test* model of the game $\mathfrak{S}(C)$ if $\mathrm{col}\, B = \mathrm{col}\, C$ and the sets of allowed moves at corresponding positions in the two games satisfy the following conditions:

$$S(C) \subset S(B),$$

$$\left.\begin{array}{r} S(\mathrm{fin}(C, \Lambda)) \cap M(\mathrm{fin}(B, \Lambda)) \subset \\ \subset S(\mathrm{fin}(B, \Lambda)) | \mathrm{col}\, \Lambda = \mathrm{col}\, C; \\ S(\mathrm{fin}(B, \Lambda)) \cap M(\mathrm{fin}(C, \Lambda)) \subset \\ \subset S(\mathrm{fin}(C, \Lambda)) | \mathrm{col}\, \Lambda \neq \mathrm{col}\, C \end{array}\right\} \Lambda \diamondsuit \mathfrak{S}(C), \Lambda \diamondsuit \mathfrak{S}_{\mathrm{T}}(B)$$

If the branch Λ is incident on both subtrees $\mathfrak{S}_{\mathrm{T}}(B)$ and $\mathfrak{S}(C)$, we shall say that the position $\mathrm{fin}(B, \Lambda)$ ($\mathrm{fin}(C, \Lambda)$) is *obtained from* $\mathrm{fin}(C, \Lambda)$ ($\mathrm{fin}(B, \Lambda)$) *by parallel transfer*.

If $\mathrm{col}\, B = \mathrm{col}\, C$ and $S(C) \subset M(B)$ we may construct the test model $\mathfrak{S}_{\mathrm{T}}(B)$ for a given game $\mathfrak{S}(C)$ recursively, defining the moves allowed at the test positions as follows:

$$S(B) := S(C);$$
$$S(\mathrm{fin}(B, \Lambda)) := M(\mathrm{fin}(B, \Lambda)) \cap$$
$$\cap\, S(\mathrm{fin}(C, \Lambda)) | \Lambda \diamond \mathfrak{S}(C),\ \Lambda \diamond \mathfrak{S}_{\mathrm{T}}(B).$$

Normally, however, we do not know the game $\mathfrak{S}(C)$ nor even the position C itself when we specify the base B of the test model $\mathfrak{S}_{\mathrm{T}}(B)$. If, however, we do know that the sets of allowed moves at the positions of $\mathfrak{S}(C)$ must belong to the subsets of virtual moves $R(\Lambda) \subset V$ defined by the branches Λ leading to those positions, we may attempt to construct the test model $\mathfrak{S}_{\mathrm{T}}(B)$.

The test game is defined recursively by its sets of allowable moves:

$$S(B) := M(B) \cap R(\varnothing);$$
$$S(\mathrm{fin}(B, \Lambda)) := M(\mathrm{fin}(B, \Lambda)) \cap R(\Lambda) | \Lambda \diamond \mathfrak{S}_{\mathrm{T}}(B).$$

Here $R(\varnothing)$ is the set of virtual moves to which the set $S(C)$ of moves allowed at the base C of $\mathfrak{S}(C)$ must belong. In some model games used in algorithms for game programming the subsets $R(\Lambda)$ of the set V of all virtual moves coincide for all the subtrees $\mathfrak{A}(C)$ for positions C of sufficiently high rank. In the extreme case, $R(\Lambda) \equiv V$. If B is of lower rank, the model $\mathfrak{S}_{\mathrm{T}}(B)$ with base B is a test model for those games $\mathfrak{A}(C)$ in which virtual moves that do not exist at B are disallowed at the base C.

When $\mathfrak{S}_{\mathrm{T}}(B)$ is a test model for a number of the games $\mathfrak{A}(C_i)$ ($i = 1, 2, \ldots, k$) there is a potential for shortening the search of the original game $\mathfrak{A}$, as a consequence of a theorem on the transfer of scores of moves which we now state and prove.

Theorem on the Transfer of Scores of Moves. *Let the following conditions be satisfied*:

(1) *In the game* $\mathfrak{A}$: *White and Black move alternately*; *a set* V *of virtual moves is defined, as is an influence relation of branches on moves, branches, and trees that satisfies the axioms and definitions given in this section*;
(2) $\mathfrak{S}(C) \subset \mathfrak{A}$ *is a model game with base* C *in which an arbitrary move* Θ *that is admissible under the rules of* $\mathfrak{A}$ *at* C *but not admissible there under the rules of* $\mathfrak{S}(C)$ *is inadmissible at the other positions of* col C *in the model game*;
(3) L *is a branch of even length, beginning at* B *and ending at* C, *and* $\mathfrak{S}_{\mathrm{T}}(B)$ *is a test model game for* $\mathfrak{S}(C)$ *with base* B;
(4) $\Psi_1 \in S(C) \subset S(B)$ *is a virtual move*, $\Psi_1 \in (B, B_1) = (C, C_1)$. $\mathfrak{S}_{\mathrm{T}}(B, \Psi_1)$ *and* $\mathfrak{S}(C, \Psi_1)$ *are subtrees of* $\mathfrak{S}_{\mathrm{T}}(B)$ *and* $\mathfrak{S}(C)$ *respectively, consisting of positions and arcs in the branches* $K(\Psi_1, \ldots)$ *having* Ψ_1 *as first move* (see Figure 27).

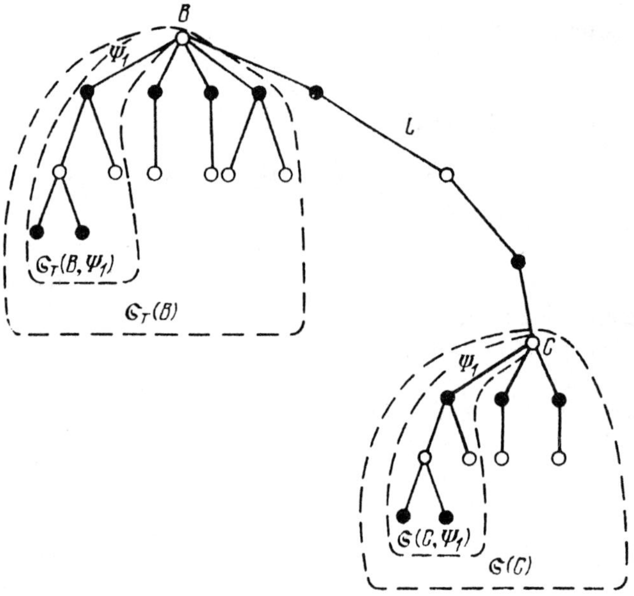

Figure 27

(5) *The position* B_1 *has the score* $\mathrm{sc}_{\mathfrak{S}_\mathrm{T}(B)}(B_1) \underset{\mathrm{col}\,C}{\prec} m$ *in the game* $\mathfrak{S}_\mathrm{T}(B)$. $\mathfrak{S}'_\mathrm{T}(B)$ *is a minimal* col B_1*-pruned tree of the game* $\mathfrak{S}_\mathrm{T}(B, \Psi_1)$ *for whose every terminal position F we have* $\mathrm{sc}_{\mathfrak{S}_\mathrm{T}(F)}(B_1) \underset{\mathrm{col}\,C}{\prec}$.

Then either the position C has a score $\mathrm{sc}_{\mathfrak{S}(C)}(C) \underset{\mathrm{col}\,C}{\prec} m$ *in the game* $\mathfrak{S}(C)$, *or the branch L influences the tree* $\mathfrak{S}'_\mathrm{T}(B, \Psi_1)$, *or it influences some move* Θ *admissible under the rules of the game* $\mathfrak{A}$ *and allowed at the positions B and C but either inadmissible or disallowed in the game* $\mathfrak{S}_\mathrm{T}(B)$ *after the move* Ψ_1 *and some reply to it* Ψ_2.

We may exclude the axioms on Cases 3, 4, and 5 from our axiom system. Then we must add the following item to the hypothesis of our theorem:

(6) *If K is a branch incident on the trees* $\mathfrak{S}_\mathrm{T}(B)$ *and* $\mathfrak{S}(C)$, *and the positions* $\mathrm{fin}(B, K)$ *and* $\mathrm{fin}(C, K)$ *are terminal, then*

$$\mathrm{sc}_{\mathfrak{S}(C)}(\mathrm{fin}(C, \Lambda)) \underset{\mathrm{col}\,\Lambda}{\preccurlyeq} \mathrm{sc}_{\mathfrak{S}_\mathrm{T}(B)}(\mathrm{fin}(B, \Lambda));$$

if, however, only one of these positions is terminal, then

$$\mathrm{sc}_{\mathfrak{S}_\mathrm{T}(B)}(\mathrm{fin}(B, \Lambda)) \underset{\mathrm{col}\,\Lambda}{\prec} m \mid S(\mathrm{fin}(B, \Lambda)) = \varnothing,$$

$$\mathrm{sc}_{\mathfrak{S}(C)}(\mathrm{fin}(C, \Lambda)) \underset{\mathrm{col}\,\Lambda}{\prec} m \mid S(\mathrm{fin}(C, \Lambda)) = \varnothing,$$

From now on we shall say that a position D in the trees $\mathfrak{S}_\mathrm{T}(B)$ and $\mathfrak{S}(C)$ is a *losing, or lost, position* if it has scores in the corresponding games that are worse than m for the color col(C), and moves that lead to a losing position will be called *losing moves*.

The basic ideas of the proof were set out in the preceding section. We repeat them here and mark the fundamental steps in the proof, which in an altered form appears in the proofs of other theorems on the possibility of shortening the search in the game tree by using the influence relation.

The first step is the construction of the branch Λ, which is similar to the critical branch described in Chapter 1. The latter is the intersection of minimal W-pruned and B-pruned trees coordinated with respect to scores.

We take the intersection of the trees $\mathfrak{S}'_T(B,\Psi_1)$ and $\mathfrak{S}''(C,\Psi_1)$. We recall that $\mathfrak{S}'_T(B,\Psi_1)$ is a minimal pruned subtree of the game tree $\mathfrak{S}_T(B,\Psi_1)$ of the color opposite to col $B=$ col C; the terminal positions of the intersecting trees have scores $\underset{\operatorname{col} C}{\prec}$ m, by the hypothesis of the theorem; $\mathfrak{S}''(C,\Psi_1)$ is a minimal col C-pruned subtree of the game $\mathfrak{S}(C,\Psi_1)$, in which it defines the score of the position C. Although these trees have different roots, their intersection can be defined by parallel transfer of branches.

If the position C_1 is lost in the game $\mathfrak{S}(C)$, one of the alternatives in our theorem is satisfied. We therefore assume that it is not lost. Then no move $(D,E)\in\mathfrak{S}''(C,\Psi_1)$ loses, and no terminal position of $\mathfrak{S}''(C,\Psi_1)$ is lost.

The first move in the branch Λ is $\Psi_1=(B,B_1)=(C,C_1)$. If B_1 is non-terminal in the game $\mathfrak{S}_T(B)$, the tree $\mathfrak{S}'_T(B,\Psi_1)$ contains a move Ψ_2 leading from B_1 to a lost position $B_2\in\mathfrak{S}'_T(B,\Psi_1)$. Such a move is unique since $\mathfrak{S}'_T(B,\Psi_1)$ is minimal. Suppose that it is admissible under the rules of the game $\mathfrak{A}$ and allowed under the rules of the model game $\mathfrak{S}(C)$ at the position $C_1\in\mathfrak{S}''(C,\Psi_1)$ obtained from B_1 by parallel transfer. Since the color of the move $\Psi_2=(C_1,C_2)$ is opposite to col C, both the move and the position C_2 to which it leads belong to the col C-pruned tree $\mathfrak{S}''(C,\Psi_1)$.

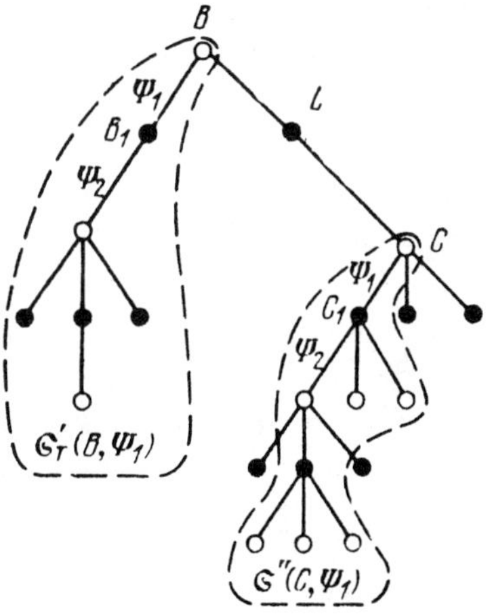

Figure 28

In the same way we transfer to the tree $\mathfrak{S}''(B, \Psi_1)$ the unique non-losing move $\Psi_3 = (C_2, C_3) \in \mathfrak{S}''(C, \Psi_1)$ if it exists, and so on. (See Figure 28.)

Thus we have uniquely defined the branch Λ incident on the trees $\mathfrak{S}'_{\mathrm{T}}(B,\Psi_1)$ and $\mathfrak{S}''(C, \Psi_1)$ and not further extendable. It is the intersection of $\mathfrak{S}'_{\mathrm{T}}(B, \Psi_1)$ and the tree $\mathfrak{S}''(C, \Psi_1)$ obtained by parallel transfer of the base B of the former. We call the branch Λ *pseudocritical.*

In the second step of the proof we see why the branch Λ cannot be extended. There may be two reasons:

(1) the branch Λ as constructed from B or C may have reached a terminal position in the corresponding model game;
(2) the next move Ψ_k chosen at one of the branch ends $\mathrm{fin}(B, \Lambda)$ or $\mathrm{fin}(C, \Lambda)$ is not allowed under the rules of the game $\mathfrak{A}$ or is inadmissible in the other branch.

As we showed in the preceding section, five different cases may arise at the ends $\mathrm{fin}(B, \Lambda)$ and $\mathrm{fin}(C, \Lambda)$.

We now go to the third step in the proof—the investigation of each of these cases and the proof that the branch Λ influences some branch of the tree $\mathfrak{S}'_{\mathrm{T}}(B, \Psi_1)$ or influences a move satisfying the conditions of the theorem.

(1) the branch Λ has odd length $2j-1\,(j \geq 1)$ and the move Ψ_{2j} at its end (the position $B_{2j-1} = \mathrm{fin}(B, \Lambda) \in \mathfrak{S}'_{\mathrm{T}}(B, \Psi_1)$) that would extend it in the tree $\mathfrak{S}'_{\mathrm{T}}(B, \Psi_1)$ is inadmissible under the rules of the game $\mathfrak{A}$ at the corresponding position $C_{2j-1} = \mathrm{fin}(C, \Lambda) \in \mathfrak{S}''(C, \Psi_1)$ or is not allowed in the model game $\mathfrak{S}(C)$. In particular, this case arises when the position C_{2j-1} is terminal and B_{2j-1} is not. Since $\mathrm{col}\, C_{2j-1} \neq \mathrm{col}\, C$, moves that are allowed at B_{2j-1} and admissible at C_{2j-1} are allowed there. Of course, the move Ψ_{2j} would be allowed at C_{2j-1} if it were admissible there, i.e. it is inadmissible under the rules of the game $\mathfrak{A}$. By the axiom on Case 1, $L \sim \Lambda * \Psi_{2j}$. But the branch $\Lambda * \Psi_{2j}$ is incident on the tree $\mathfrak{S}'_{\mathrm{T}}(B, \Psi_1)$ and therefore $L \sim \mathfrak{S}'_{\mathrm{T}}(B, \Psi_1)$.
(2) The branch Λ has even length $2j\,(j \geq 1)$ and the move Ψ_{2j+1} at its end

$$C_{2j} = \mathrm{fin}(C_1, \Lambda) \in \mathfrak{S}''(C, \Psi_1),$$

which would extend it in the tree $\mathfrak{S}''(C, \Psi_1)$, is either inadmissible under the rules of the game $\mathfrak{A}$ or is not allowed in the game $\mathfrak{S}(C)$ at the position

$$B_{2j} = \mathrm{fin}(B, \Lambda) \in \mathfrak{S}'_{\mathrm{T}}(B, \Psi_1).$$

Then B_{2j} also may be a terminal position. Since $\mathrm{col}\, C_{2j} = \mathrm{col}\, C$, moves allowed at C_{2j} and admissible at B_{2j} are allowed there. Accordingly the move Ψ_{2j+1} is inadmissible under the rules of the game $\mathfrak{A}$. If it is not admissible at $C = \mathrm{fin}(B, L)$, the first axiom on Case 2 implies that $L \sim \Lambda$ and so $L \sim \mathfrak{S}'_{\mathrm{T}}(B, \Psi_1)$. If, however, it is admissible at C it is allowed there, by the conditions of the theorem (else it would not be allowed at $C_{2j} = \mathrm{fin}(C, \Lambda) \in \mathfrak{S}(C)$). It is also allowed at B, since $S(C) \subset S(B)$. By the second axiom on Case 2, $L \sim \Lambda$ or $L \sim \Psi_{2j+1}$. In the first alternative, as before, $L \sim \mathfrak{S}'_{\mathrm{T}}(B, \Psi_1)$.

In the second alternative, if the move Ψ_{2j+1} is admissible and allowed at the position $B_2 \in \mathfrak{S}'_{\mathrm{T}}(B, \Psi_1)$ after the moves $\Psi_1 = (B, B_1)$ from B and $\Psi_2 = (B_1, B_2)$ from B_1, the move $\Psi_{2j+1} = (B_{2j}, D)$ and the position D belong to the col B_1-pruned tree $\mathfrak{S}'_{\mathrm{T}}(B, \Psi_1)$ since

$$\mathrm{col}\,\Psi_{2j+1} = \mathrm{col}\,C_{2j} = \mathrm{col}\,C = \mathrm{col}\,B \neq \mathrm{col}\,B_1.$$

Accordingly, the branch $\Lambda'(\Psi_1, \Psi_2, \ldots, \Psi_{2j+1})$ is incident on the tree $\mathfrak{S}'_{\mathrm{T}}(B, \Psi_1)$. By the axiom on influences on moves and branches, $L \sim \Lambda'$ and so $L \sim \mathfrak{S}'_{\mathrm{T}}(B, \Psi_1)$. Now there remains only the case in which the branch L influences the move Ψ_{2j+1}, which is admissible and allowed at the beginning B and end C of the branch, but not admissible or not allowed after a move from B and the reply Ψ_2. In concrete games we may formulate supplementary conditions to be satisfied by the tree $\mathfrak{S}'_{\mathrm{T}}(B, \Psi_1)$ in order that the move Ψ_{2j+1} be distinct from Ψ_1.

(3) The branch Λ has odd length $2j-1\,(j \geq 1)$, its end

$$B_{2j-1} = \mathrm{fin}(B, \Lambda) \in \mathfrak{S}'_T(B, \Psi_1)$$

is terminal in $\mathfrak{S}_{\mathrm{T}}(B)$, and the corresponding position

$$C_{2j-1} = \mathrm{fin}(C, \Lambda) \in \mathfrak{S}''(C, \Psi_1)$$

is non-terminal in the game tree $\mathfrak{S}(C)$. Then B_{2j-1} is a terminal position even in the original game $\mathfrak{A}$, since the move at it belongs to the color opposite to col B. Moreover, if $M(B_{2j-1}) \neq \varnothing$ and $S(B_{2j-1}) = \varnothing$, that color loses. By the axiom on Cases 3 and 4, $L \sim \Lambda$, and so $L \sim \mathfrak{S}'_{\mathrm{T}}(B, \Psi_1)$. If we renounce this axiom and add the condition in item 6, the situation we have in hand is impossible. In fact, since $\mathrm{col}\,\Lambda \neq \mathrm{col}\,B$, the condition

$$\mathrm{sc}_{\mathfrak{S}_T(B)}(\mathrm{fin}(B, \Lambda)) \underset{\mathrm{col}\,\Lambda}{\prec} m$$

implies that

$$m \underset{\mathrm{col}\,B = \mathrm{col}\,C}{\prec} \mathrm{sc}_{\mathfrak{S}_{\mathrm{T}}(B)}(\mathrm{fin}(B, \Lambda)),$$

and this contradicts the assertion of Condition 5 that the position $B_{2j} = \mathrm{fin}(B, \Lambda)$ is lost.

(4) The branch Λ has even length $2j\,(j \geq 1)$, its end $C_{2j} = \mathrm{fin}(C, \Lambda) \in \mathfrak{S}''(C, \Psi_1)$ is a terminal position in the game $\mathfrak{S}(C)$ and the corresponding position $B_{2j} = \mathrm{fin}(B, \Lambda) \in \mathfrak{S}'_{\mathrm{T}}(B, \Psi_1)$ is non-terminal in the game $\mathfrak{S}_{\mathrm{T}}(B)$. Then we prove the following facts in exactly the same way as above: the position C_{2j} is terminal in the game $\mathfrak{A}$; if the axiom on Cases 3 and 4 holds we have $L \sim \Lambda$; if Condition 6 holds C_{2j} is a lost position.

(5) Both of the positions $\operatorname{fin}(B, \Lambda)$ and $\operatorname{fin}(C, \Lambda)$ are terminal in the corresponding model games, but both cannot be non-terminal in the original game $\mathfrak{A}$, since then the rule on the definition of scores would imply that both were lost or both were not lost. If one of them is terminal in the model game and the other is terminal in the original game, the axiom on Cases 3 and 4 implies that $L \sim \Lambda$. Finally, if both positions are terminal in the original game $\mathfrak{A}$ the axiom on Case 5 implies that

$$\operatorname{sc}(\operatorname{fin}(C, \Lambda)) \underset{\operatorname{col} C}{\prec} \operatorname{sc}(\operatorname{fin}(B, \Lambda))$$

or $L \sim \Lambda$. But Condition 5 implies that the conditions and our postulates for the first possibility are not realized. Thus $L \sim \mathfrak{S}'_{\mathrm{T}}(B, \Psi_1)$. If we renounce the axiom on Case 5 and add the condition of item 6, this situation too will be impossible, since the requirements that the position $\operatorname{fin}(B, \Lambda)$ be lost and $\operatorname{fin}(C, \Lambda)$ not lost and

$$\operatorname{sc}_{\mathfrak{S}(C)}(\operatorname{fin}(C, \Lambda)) \underset{\operatorname{col} C}{\preccurlyeq} \operatorname{sc}_{\mathfrak{S}_{\mathrm{T}}(B)} \operatorname{fin}(B, \Lambda)$$

are incompatible.

This completes the proof of the theorem on the transfer of scores. Some of the axioms were not used in the proof, in particular, the axiom on the symmetry of the influence relation. If we define the influence relation constructively in such a way that we use only the above theorem and the following theorem on the transfer of position scores, we do not require symmetry and we construct a relationship that will be more rarely satisfied and so will offer more frequent possibilities for shortening the search.

The following theorem is proved in exactly the same way as the above.

Theorem on the Transfer of Position Scores. *Let the following conditions be satisfied*:

(a) *conditions* 1–3 *of the theorem on the transfer of scores of moves*;

(b) *the position B has the following score in the game* $\mathfrak{S}_{\mathrm{T}}(B)$

$$\operatorname{sc}_{\mathfrak{S}_{\mathrm{T}}(B)}(B) \underset{\operatorname{col} C}{\prec} m,$$

(c) $\mathfrak{S}'_{\mathrm{T}}(B)$ *is a minimal pruned tree of the game* $\mathfrak{S}_{\mathrm{T}}(B)$ *for the color opposite to* $\operatorname{col} B = \operatorname{col} C$ *and all its terminal positions F have a score*

$$\operatorname{sc}_{\mathfrak{S}_{\mathrm{T}}(B)}(F) \underset{\operatorname{col} C}{\prec} m.$$

Then the position C has a score

$$\operatorname{sc}_{\mathfrak{S}(C)}(C) \underset{\operatorname{col} C}{\prec} m$$

in the game $\mathfrak{S}(C)$ *else the branch L influences the tree* $\mathfrak{S}'_{\mathrm{T}}(B)$.

As in the theorem on the transfer of scores of moves, we may impose the condition 6 instead of the axiom on Cases 3, 4, and 5.

Let us ask why we discuss the influence of the branch L on the tree $\mathfrak{S}'_T(B)$ or the tree $\mathfrak{S}'_T(B,\Psi_1)$ but not on the pseudocritical branch Λ. When we arrive at the position C in the search of the game tree $\mathfrak{A}$ or one of its models, the search of $\mathfrak{S}'_T(B)$ is normally already completed, so that we already know both it and its subtree $\mathfrak{S}'_T(B,\Psi)$, where $\Psi \in S(B)$, but the C-subtree of $\mathfrak{A}$ or a model of $\mathfrak{A}$ has not yet been searched. Therefore we do not then know the pseudocritical branch Λ but must make a decision about shortening the search. For this reason we must use information about the branch $L(B,\ldots,C)$ and the trees $\mathfrak{S}'_T(B)$ and $\mathfrak{S}'_T(C,\Psi_1)$ for $\Psi \in S(B)$.

The Decomposition of Branches and Application to Shortening the Search

One of the unpleasant constraints in the hypothesis of the theorem on the transfer of move scores is that at the position C, where we wish to shorten the search, we cannot contemplate any move Θ that is inadmissible under the rules of the original game $\mathfrak{A}$ at the base B of the test model tree $\mathfrak{S}_T(B)$. Thus from the point of view of col B the position B should 'strongly majorize' the position C. However, as a rule the moves in the branch L beginning at B and ending at C offer the opponent new possibilities that often bear no relation to the variants of the model game $\mathfrak{S}_T(B)$.

On the other hand, it was assumed in the proof of the theorem that the tree $\mathfrak{S}_T(B)$ was searched without pruning, because some of its branches do not influence others. Thus the search takes more time than may be necessary. It is clearly not easy to investigate the possibility of shortening the search by recursive pruning based on repeated application of the method of analogy. Only the first steps have been made in this direction. They require new ideas, which will probably be of a rather general character. In this section we attempt to present some thoughts about them. We consider certain possibilities for shortening the search when, at the end $C = \mathrm{fin}(B, L)$ of a branch L having even length, certain moves Θ are admissible even though they do not belong to the set $S(B)$ of moves allowed at the base position of the test model game $\mathfrak{S}_T(B)$. Some of these possibilities are connected with the permissibility of studying some of the moves $\Psi \in S(B)$ independently of one another.

Let $\mathfrak{S}$ be a model game with the following properties:

At the positions $\mathrm{fin}(C,\Lambda)$ of its subtrees $\mathfrak{S}(C)$ all virtual moves in the subsets $R(\Lambda)$, and only those, are allowed, i. e.

$$S(\mathrm{fin}(C,\Lambda)) = M(\mathrm{fin}(C,\Lambda)) \cap R(\Lambda) \mid \Lambda \diamondsuit \mathfrak{S}(C).$$

The set $R(\varnothing) \subset V$ is also defined, corresponding to the empty branch $\varnothing$:

$$S(C) = M(C) \cap R(\varnothing).$$

The sets decrease as new moves are added to the branch Λ:

$$R(\Lambda_1+\Lambda_2)\subset R(\Lambda_1)\,|\mathrm{col}(\Lambda_1+\Lambda_2)=\mathrm{col}\,\Lambda_1.$$

In particular, any set $R(\Lambda)\subset R(\varnothing)$.

Many model games satisfy such conditions, for instance, the absolute scheme

$$R(\Lambda)=\begin{cases} V, & \text{if } r(\mathrm{fin}(C,\Lambda))<n,\\ V_f, & \text{if } r(\mathrm{fin}(C,\Lambda))\geq n,\end{cases}$$

where $r(D)$ is the rank of the position D in the tree $\mathfrak{A}$, n is the depth of the model, and $V_f\subset V\cup m_\varnothing$ is the set of moves in the forced game. For a quiet game

$$R(\Lambda)=\begin{cases} V, & \text{if } r(\mathrm{fin}(C,\Lambda))<n\\ & \text{or}\,|V_f\cap\{\Lambda\}|>\rho(\Lambda)-d(C),\\ V_f & \text{in other cases,}\end{cases}$$

where $\{\Lambda\}$ is the set of moves in the branch Λ, $\rho(\Lambda)$ is its length, $|U|$ is the number of elements in the set U, and $d(C)$ is the number of quiet moves allowed in the branch. Finally, the original game $\mathfrak{A}$ itself meets the conditions: we need only assume that $R(\Lambda)=V$ for all Λ.

The test model game $\mathfrak{S}(B)$ that we shall be considering later must satisfy the following conditions:

(1) An arbitrary move $\Psi\in S(B)$ from the base position loses, that is

$$\mathrm{sc}_{\mathfrak{S}_T(B)}(B)\underset{\mathrm{col}\,B}{\prec} m,$$

where m is a given score.

(2) A new move Θ from the position

$$D=\mathrm{fin}(B,\Lambda)\in\mathfrak{S}_1(B)$$

of the color $\mathrm{col}(B)$, i.e. a move not existing at B, is allowed if it belongs to the corresponding set $R(\Lambda)$:

$$M(\mathrm{fin}(B,\Lambda))\setminus M(B)\cap R(\Lambda)\subset S(\mathrm{fin}(B,\Lambda))$$

for $\mathrm{col}\,\Lambda=\mathrm{col}\,B$.

(3) If a branch Λ of color $\mathrm{col}\,\Lambda=\mathrm{col}\,B$ is incident on the tree of the game $\mathfrak{S}_T(B)$, all the moves $\Psi\in M(\mathrm{fin}(B,\Lambda))\cap M(B)\cap R(\Lambda)$ that are influenced by Λ are allowed at the position $\mathrm{fin}(B,\Lambda)$:

$$\Psi\in M(\mathrm{fin}(B,\Lambda))\cap M(B)\cap R(\Lambda)\,\&\,\Lambda\sim\Psi$$

$$\mathrm{col}\,\Lambda=\mathrm{col}\,B\Rightarrow\Psi\in S(\mathrm{fin}(B,\Lambda)).$$

(4) Let

(a) $\mathfrak{S}'_{\mathrm{T}}(B)$ be a minimal pruned tree of the game $\mathfrak{S}_{\mathrm{T}}(B)$, of color opposite to $\mathrm{col}(B)$;

(b) the terminal positions $F \in \mathfrak{S}'_{\mathrm{T}}(B)$ have scores $\mathrm{sc}_{\mathfrak{S}(B)}(F) \underset{col\,B}{\prec} \mathrm{m}$;

(c) $\mathfrak{S}_{\mathrm{T}}(B, \Psi)$ be a subtree of $\mathfrak{S}_{\mathrm{T}}(B)$, consisting of the positions and moves in the branch $\Lambda(\Psi, \dots) \diamondsuit \mathfrak{S}_{\mathrm{T}}(B)$, of which the first move init Λ is $\Psi \in S(B)$;

(d) $\mathfrak{S}'_{\mathrm{T}}(B, \Psi) = \mathfrak{S}'_{\mathrm{T}}(B) \cap \mathfrak{S}_{\mathrm{T}}(B, \Psi)$.

Then if the tree $\mathfrak{S}'_{\mathrm{T}}(B, \Psi_1)$ influences the tree $\mathfrak{S}'_{\mathrm{T}}(B, \Psi_2)$ ($\Psi_1, \Psi_2 \in S(B)$), the move Ψ_2 is allowed at all positions of the tree $\mathfrak{S}_{\mathrm{T}}(B, \Psi_1)$ where the side col B has the move and Ψ_2 is admissible under the rules of $\mathfrak{A}$ and belongs to the corresponding set $B(\Lambda)$:

$$\mathfrak{S}'_{\mathrm{T}}(B, \Psi_1) \sim \mathfrak{S}'_{\mathrm{T}}(B, \Psi_2) \& \Lambda \diamondsuit \mathfrak{S}_{\mathrm{T}}(B, \Psi_1) \&$$

$$\Psi_2 \in M(\mathrm{fin}(B, \Lambda)) \cap R(\Lambda) \Rightarrow \Psi_2 \in S(\mathrm{fin}(B, \Lambda)).$$

In particular, the move Ψ_1 is allowed at all the positions $\mathrm{fin}(B, \Lambda) \in \mathfrak{S}_{\mathrm{T}}(B, \Psi_1)$ where it is admissible and belongs to the corresponding set $R(\Lambda)$.

By the axiom on the symmetry of the influence relation the move Ψ_1 is also allowed at the positions in $\mathfrak{S}_{\mathrm{T}}(B, \Psi_2)$ if it is admissible and belongs to the corresponding sets $R(\Lambda)$.

(5) All the moves allowed at the position $D = \mathrm{fin}(B, \Lambda) \in \mathfrak{S}_{\mathrm{T}}(B)$ where the color opposite to col B has the move, belong to the set $R(\Lambda)$:

$$S(\mathrm{fin}(B, \Lambda)) \subset M(\mathrm{fin}(B, \Lambda)) \cap$$

$$R(\Lambda) \mid \Lambda \diamondsuit \mathfrak{S}_{\mathrm{T}}(B), \mathrm{col}\,\Lambda \neq \mathrm{col}\,B.$$

Thus we want to study individually only those moves Ψ that are allowed at the base B.

It has been conjectured that at any position one need investigate only new moves Θ and those earlier possible moves whose branches are influenced by the search tree already constructed there. A proof of this conjecture requires an extended concept of influence on branches with different origins, and has not so far been obtained. Its application entails some algorithmic difficulties, since a return may be required to positions from which a backward step has already been taken.

We can construct by iteration a model game $\mathfrak{S}_{\mathrm{T}}(B)$ in which, at positions of color col B that do not coincide with the base B, moves admissible at B are not allowed if possible. Let $G_0 := S(B)$ and let the set $H_0(\mathfrak{J})$ be empty for all $\Psi \in S(B)$. Let the sets $G_i \subset S(B)$ and $H_i(\Psi)$ for $\Psi_i \in G_i$ be defined for some $i \geq 0$. Then we define the model game $\mathfrak{S}_i(B, \Psi)$

recursively for all $\Psi \in G_i$:

$$S_\Psi(B) := \{\Psi\};$$

$$S_\Psi(\mathrm{fin}(B,\Lambda)) :=$$

$$:= \begin{cases} ((M(\mathrm{fin}(B,\Lambda)) \setminus M(B)) \cup (M(\mathrm{fin}(B,\Lambda)) \cap H_i(\Psi))) \\ \qquad \cap R(\Lambda) | \mathrm{col}\,\Lambda = \mathrm{col}\,B; \\ M(\mathrm{fin}(B,\Lambda)) \cap R(\Lambda) | \mathrm{col}\,\Lambda \neq \mathrm{col}\,B. \end{cases}$$

Let us now see whether the position B is lost in these games, i. e. whether its score $\underset{\mathrm{col}\,B}{\prec} m$. To this end we construct for the losing moves Ψ the corresponding minimal pruned trees $\mathfrak{S}'_i(B,\Psi)$ of the color opposite to col B. Then

$$G_{i+1} := \left\{\Psi \mid \Psi \in G_i \,\&\, \mathrm{sc}_{\mathfrak{S}_i(B,\Psi)}(B) \underset{\mathrm{col}\,B}{\prec} m\right\};$$

$$H_{i+1}(\Psi) := G_{i+1} \cap (H_i(\Psi) \cup \{\Theta \mid \mathfrak{S}'_i(B,\Theta) \sim \mathfrak{S}'_i(B,\Theta)\}).$$

In the latter definition we assume that the influence relation is symmetric, else we would need to add the moves Θ for which $\mathfrak{S}'_i(B,\Theta) \sim \mathfrak{S}'_i(B,\Psi)$. Then we may go to the following iteration.

Since $G_0 = S(B)$ is finite and G_{i+1} is imbedded in G_i for all i greater than some N, these sets will stabilize:

$$i > N \Rightarrow G_i = G_N.$$

In the following iterations the sets $H_i(\Psi)$ can only expand, but they remain subsets of the finite set G_N, so that the expansion process must end. So for some $i > N$ we shall have the following conditions:

$$G_{i+1} = G_i = G_N,$$

$$\forall \Psi \in G_i \quad H_{i+1}(\Psi) = H_i(\Psi).$$

Hence the model games $\mathfrak{S}_{i+1}(B,\Psi)$ and $\mathfrak{S}_i(B,\Psi)$ will be identical.

The model game

$$\mathfrak{S}_{\mathrm{T}}(B) = \bigcup_{\Psi \in G_i} \mathfrak{S}_i(B,\Psi)$$

satisfies the conditions we have imposed on a test model, and $\mathfrak{S}_i(B,\Psi)$ and $\mathfrak{S}'_i(B,\Psi)$ ($\Psi \in G_i = S_{\mathrm{T}}(B)$) are the subtrees of $\mathfrak{S}_{\mathrm{T}}(B,\Psi)$ and $\mathfrak{S}'_{\mathrm{T}}(B,\Psi)$, respectively, that we discussed in item 3 of the conditions.

In fact, the following three assertions are valid:

(1) All the moves $\Psi \in G_i = G_{i+1}$ are losing moves.

(2) Let

$$\mathfrak{S}_{\mathrm{T}}(B) = \bigcup_{\Psi \in G_i} \mathfrak{S}_i(B,\Psi) = \bigcup_{\Psi \in G_{i+1}} \mathfrak{S}_{i+1}(B,\Psi),$$

where Λ is a branch incident on the tree $\mathfrak{S}_T(B)$. If col Λ = col B,

$$(M(\text{fin}(B,\Lambda))\setminus M(B))\cap R(\Lambda)\subset S(\text{fin}(B,\Lambda))$$

since by definition

$$S(\text{fin}(B,\Lambda)) = ((M(\text{fin}(B,\Lambda))\setminus M(B)$$
$$\cup(M(\text{fin}(B,\Lambda))\cap H_{i+1}(\Psi)))\cap R(\Lambda)$$

If however, col $\Lambda \neq$ col B we have

$$M(\text{fin}(B,\Lambda))\cap R(\Lambda) = S(\text{fin}(B,\Lambda)).$$

(3) If $\mathfrak{S}'_T(B,\Psi_1) \sim \mathfrak{S}'_T(B,\Psi_2)$ $(\Psi_1,\Psi_2\in S(B),\ G_i = G_{i+1})$, then

$$\Psi_1\in H_i(\Psi_2) = H_{i+1}(\Psi_2),$$
$$\Psi_2\in H_i(\Psi_1) = H_{i+1}(\Psi_1).$$

Thus if the move Ψ_2 is admissible at the position $\text{fin}(B,\Lambda)\in\mathfrak{S}_T(B,\Psi_1)$ and belongs to the corresponding set, it is allowed. A like assertion is valid also for the move Ψ_1 at positions of color col B in the subtree $\mathfrak{S}_T(B,\Psi_2)$.

Such a process, however, may be too lengthy. In practice, if the second or third iteration alters the sets G_i or $H_i(\Psi)$ we should obviously construct a model with more moves but a simpler rule, as follows:

For every move $\Psi\in M(B)\cap R(\varnothing)$ or belonging to the subset we are interested in, $S(B)\subset M(B)\cap R(\varnothing)$, we define the game $\mathfrak{S}\text{T}(B,\Psi)$ recursively:

$$S_\Psi(B) = \{\Psi\};$$
$$S_\Psi(\text{fin}(B,\Lambda)) := M(\text{fin}(B,\Lambda))\cap R(\Lambda)$$
$$|\Lambda\neq\varnothing,\ \Lambda\diamondsuit S_T(B,\Psi).$$

Next we may adjust m so that

$$\text{sc}_{\mathfrak{S}_T(B,\Psi)}(B) \underset{\text{col } B}{\prec} m \mid \Psi\in S(B),$$

and look at the earlier defined set $S(B)\subset M(B)$ of moves allowed at the base B, or if m is fixed, adjust the set $S(B)$:

$$S(B) := \left\{\Psi \mid \Psi\subset S(B)\&\text{sc}_{\mathfrak{S}_T(B,\Psi)}(B) \underset{\text{col } B}{\prec} m\right\}.$$

In both cases the test model game $\mathfrak{S}_T(B)$ is defined as the union of the games $\mathfrak{S}_T(B,\Psi)$ for $\Psi\in S(B)$:

$$\mathfrak{S}_T(B) := \bigcup_{\Psi\in S(B)} \mathfrak{S}_T(B,\Psi).$$

Theorem on the Transfer of Scores of Groups of Moves. *Let the following conditions be satisfied*:

(1) *In the game* $\mathfrak{A}$ *White and Black move alternately and there are defined a set* V *of virtual moves and influence relations of a branch on a move, branch, and subtree, and of a subtree on a subtree, satisfying the axioms and definitions given in this section.*

(2) $\mathfrak{S}(C)$ *is a model game for* $\mathfrak{A}$ *with base* C, *and*

$$S(C) = M(C) \cap R(\varnothing),$$
$$S(\mathrm{fin}(C,\Lambda)) = M(\mathrm{fin}(C,\Lambda)) \cap R(\Lambda) \mid \Lambda \diamondsuit \mathfrak{S}(C),$$
$$R(\Lambda_1 + \Lambda_2) \subset R(\Lambda_1) \mid \Lambda \diamondsuit \mathfrak{S}(C), \mathrm{col}\,\Lambda = \mathrm{col}\,\Lambda_1 = \mathrm{col}\,C.$$

(3) L *is a branch of even length with beginning* B *and end* C, $\mathfrak{S}_{\mathrm{T}}(B)$ *is a non-empty test model game for the game* $\mathfrak{S}(C)$. *Thus the following conditions are satisfied*:

$$S(B) \neq \varnothing,$$
$$S(B) \subset S(C),$$
$$\mathrm{sc}_{\mathfrak{S}\,\mathrm{T}(B)}(B) \underset{\mathrm{col}\,B = \mathrm{col}\,C}{\dashv} m,$$

$$\left.\begin{array}{l} S(\mathrm{fin}(B,\Lambda)) \subset M(\mathrm{fin}(B,\Lambda)) \cap R(\Lambda), \\ (M(\mathrm{fin}(B,\Lambda)) \setminus M(B)) \cap R(\Lambda) \subset S(\mathrm{fin}(B,\Lambda)), \\ \Theta \in S(\mathrm{fin}(B,\Lambda)) | (\Theta \in M(\mathrm{fin}(B,\Lambda)) \cap S(B) \cap R(\Lambda) \\ \qquad \& \mathfrak{S}'_{\mathrm{T}}(B,\Theta) \sim \mathfrak{S}'_{\mathrm{T}}(B, \mathrm{init}\,\Lambda)) \\ \vee (\Theta \in M(\mathrm{fin}(B,\Lambda)) \cap M(B) \cap R(\Lambda) \& \Lambda \sim \Theta), \end{array}\right\} \begin{array}{l} \Lambda \diamondsuit \mathfrak{S}_{\mathrm{T}}(B), \\ \mathrm{col}\,\Lambda = \mathrm{col}\,B \end{array}$$

$$\begin{array}{l} M(\mathrm{fin}(C,\Lambda_1+\Lambda_2)) \cap S(\mathrm{fin}(B,\Lambda_1)) \\ \quad \subset S(\mathrm{fin}(C,\Lambda_1+\Lambda_2)) | \Lambda_1 \diamondsuit \mathfrak{S}_{\mathrm{T}}(B), \Lambda_1 + \Lambda_2 \diamondsuit \mathfrak{S}_{\mathrm{T}}(C), \\ \qquad \mathrm{col}\,\Lambda_1 = \mathrm{col}(\Lambda_1+\Lambda_2) \neq \mathrm{col}\,B. \end{array}$$

Here $\mathfrak{S}'_{\mathrm{T}}(B,\Theta) = \mathfrak{S}_{\mathrm{T}}(B,\Theta) \cap \mathfrak{S}'_{\mathrm{T}}(B)$ *and* $\mathfrak{S}'_{\mathrm{T}}(B)$ *is a minimal pruned tree of a color opposite to* $\mathrm{col}\,B$, *with terminal positions* F *for which* $\mathrm{sc}_{\mathfrak{S}_{\mathrm{T}}(B)}(F) \underset{\mathrm{col}\,B}{\prec} m$.

$\mathfrak{S}_{\mathrm{T}}(B,\Theta)$ $(\Theta \in S(B))$ *is a subtree of the game tree* $\mathfrak{S}_{\mathrm{T}}(B)$ *consisting of positions and moves belonging to the branches* $\Lambda \diamondsuit \mathfrak{S}_{\mathrm{T}}(B)$ *for which* $\mathrm{init}\,\Lambda = \Theta$.

(4) $\mathfrak{S}(C)$ is a model game with base C where

$$\tilde{S}(C) = (M(C) \cap R(\varnothing)) \setminus S(B),$$

$$S(\mathrm{fin}(C,\Lambda)) = \left\{\begin{array}{l} (M(\mathrm{fin}(C,\Lambda)) \cap R(\Lambda)) \setminus S(B) \\ \text{for } \mathrm{col}\,\Lambda = \mathrm{col}\,C, \\ M(\mathrm{fin}(C,\Lambda)) \cap R(\Lambda) \\ \text{for } \mathrm{col}\,\Lambda \neq \mathrm{col}\,C, \quad \mathrm{sc}_{\tilde{\mathfrak{S}}(C)}(C) \underset{\mathrm{col}\,C}{\dashv} m. \end{array}\right\} \Lambda \diamondsuit \tilde{\mathfrak{S}}(C),$$

$\tilde{\tilde{\mathfrak{S}}}'(C)$ is a minimal pruned tree, of color opposite to col C, *for the game $\tilde{\tilde{\mathfrak{S}}}(C)$ with terminal positions F having scores* $\mathrm{sc}_{\tilde{\tilde{\mathfrak{S}}}(C)}(F) \underset{\mathrm{col}\, C}{\prec} m$.

$\tilde{\tilde{\mathfrak{S}}}'(B) = L \cup \tilde{\tilde{\mathfrak{S}}}'(C)$, that is, it consists of the positions and moves in the branch L and the tree $\tilde{\tilde{\mathfrak{S}}}'(C)$.

Then the position C is lost in the game $\mathfrak{S}(C)$ (i. e. $\mathrm{sc}_{\mathfrak{S}(C)}(C) \underset{\mathrm{col}\, B}{\prec} m$) *else the tree $\mathfrak{S}_{\mathrm{T}}'(B)$ influences the tree $\tilde{\tilde{\mathfrak{S}}}'(B)$.*

Let us dwell on the conditions of the theorem. We could change them by not requiring that the axioms on Cases 3, 4, and 5 be satisfied, but still making use of the connection between position scores, the color of the turn to move at the positions, the material score, and the blank move we discussed earlier. Then the proof would be substantially more complicated but would involve nothing in principle new. We could also replace the game $\tilde{\tilde{\mathfrak{S}}}(C)$—whose branches issuing from the base C begin with moves $\Theta \in S(C)$ that do not belong to the set $S(B)$—by a test game $\tilde{\tilde{\mathfrak{S}}}_{\mathrm{T}}(C)$ analogous to the game $\mathfrak{S}_{\mathrm{T}}(B)$. At the inner positions in the game $\tilde{\tilde{\mathfrak{S}}}_{\mathrm{T}}(C)$ with the move belonging to the color col B the moves in the set $S(C) = M(C) \cap R(\varnothing)$ are as far as possible disallowed. However, we shall not consider such generalizations of the theorem. Moreover, for specific games and constructively defined influence relationships we must investigate cases in which other axioms are not satisfied. A complete proof is given in [15] for chess and some chess model games.

The conditions to be satisfied by the set $S(D)$ of moves allowed at the positions D in the test model game $\mathfrak{S}_{\mathrm{T}}(B)$ are in general the same as those prescribed above. The only change is in the condition on the sets $S(D)$ of moves at the positions D where the move belongs to the color opposite to col B = col C:

$$M(\mathrm{fin}(C, \Lambda_1 + \Lambda_2)) \cap S(\mathrm{fin}(B, \Lambda_1)) \subset S(\mathrm{fin}(C, \Lambda_1 + \Lambda_2)).$$

This means that some moves allowed in the model $\mathfrak{S}_{\mathrm{T}}(B)$ at positions of significantly lower rank are not to be suppressed at high-ranked positions in the model $\mathfrak{S}(C)$. But we are actually playing in the original game $\mathfrak{A}$, not in the model, and we need not fix the latter. If we allow some moves in it that are needed to advance the proof of the theorem, the model does not suffer.

In the first stage of the proof we construct the pseudocritical branch L_0 and its decomposition into the branches $\Lambda_1, \Lambda_2, \ldots, \Lambda_n, \Lambda_{n+1}$, where n is the number of moves $\Psi_1, \Psi_2, \ldots, \Psi_n$ in the set $S(B)$. Each of the branches Λ_i for $i = 1, 2, \ldots, n$ will be incident on the tree $\mathfrak{S}_{\mathrm{T}}'(B, \Psi_i)$, and the branch Λ_{n+1} will be incident on the tree $\tilde{\tilde{\mathfrak{S}}}'(C)$. Some of the branches, however will be empty, and we assume that the empty branch $\varnothing$ is incident on every tree and does not influence any branch or any move. In addition to the branch L we shall consider auxiliary branches $L_1, L_2, \ldots, L_n$. These also all begin at the position B. We shall postulate that the position C is not lost, and prove the second alternative conclusion of the theorem. We denote by

$\mathfrak{S}''(C)$ the minimal col C-pruned tree of the game $\mathfrak{S}(C)$, whose terminal positions (and therefore all other positions) are not lost. Such a tree exists, since C is not lost.

In constructing the pseudocritical branch L_0 we must satisfy the following conditions:

$$\Lambda_i \diamondsuit \mathfrak{S}'_{\mathrm{T}}(B, \Psi_i), \quad i = 1, 2, \ldots, n,$$
$$\Lambda_{n+1} \diamondsuit \tilde{\mathfrak{S}}'(C),$$
$$\Lambda_i \nsim \Lambda_j, \quad i, j = 1, 2, \ldots, n, \quad i \neq j,$$
$$\mathfrak{S}'_{\mathrm{T}}(B, \Psi_i) \nsim \mathfrak{S}'_{\mathrm{T}}(B, \Psi_j), \quad i, j = 1, 2, \ldots, n, i \neq j,$$
$$\Lambda_i, \Lambda_j \neq \varnothing,$$
$$L_{i-1} = L_i + \Lambda_i, \quad i = 1, 2, \ldots, n,$$
$$L_i = L * L_i^c,$$
$$L_i^c = \Lambda_{i+1} + \cdots + \Lambda_{n+1} \diamondsuit \mathfrak{S}(C),$$
$$\mathrm{fin}(B, L_i) = \mathrm{fin}(C, L_i^c) \in \mathfrak{S}(C), \quad i = 0, 1, \ldots, n,$$
$$\mathrm{fin}(B, L_0) \in \mathfrak{S}''(C),$$
$$\mathrm{fin}(B, L_n) \in \tilde{\mathfrak{S}}'(C).$$

(some of these conditions are immediate consequences of others, and are stated here to facilitate their use later). When the length $\rho(L_0)$ of the constructed portion of the psuedocritical branch L_0 is even, all the branches Λ_i, L_i, L_i^c must have even length as well. If $\rho(L_0)$ is odd, only one of the branches Λ_i, L_h, L_h^c has odd length for $h > i$, and the last moves in these branches are identical.

To start the construction of the pseudocritical branch L_0 we set

$$\Lambda_i := \varnothing \ (i = 1, 2, \ldots, n+1);$$
$$L_i := L;$$
$$L_i^c := \varnothing \ (i = 0, 1, \ldots, n).$$

Then all our conditions will be satisfied. Now suppose that they are satisfied when the currently constructed portion of the pseudocritical branch L_0 has even length $\rho(L_0) = \rho(L) + 2j$, where $j \geq 0$. The position $\mathrm{fin}(B, L_0) = \mathrm{fin}(C, L_0^c) \in \mathfrak{S}''(C)$, i. e. it is not lost. Either it is terminal in that game, or there is a unique and non-losing move Ω_{2j+1} from it, belonging to the minimal col C-pruned tree $\mathfrak{S}''(C) \subset \mathfrak{S}(C)$. In the first case we end the construction of L_0 and begin to analyze the reasons for its non-extendability (cf. Part 1 of the analysis, below). In the second case we attempt to add the move Ω_{2j+1} at the end of the branch L_0.

We test the admissibility of the move Ω_{2j+1} at the position $\mathrm{fin}(B, L_i)$ for $i = 1, 2, \ldots, n$. If it is admissible at these positions it is allowed. For, since $\mathrm{fin}(B, L_i) = \mathrm{fin}(C, L_i^c)$, the move belongs to the set $M(\mathrm{fin}(C, L_i^c))$. Furthermore,

$$L_0^c = \underbrace{\Lambda_1 + \cdots + \Lambda_i}_{\tilde{\Lambda}_i} + \underbrace{\Lambda_{i+1} + \cdots + \Lambda_{n+1}}_{L_i^c} = L_i^c + \tilde{\Lambda}_i,$$

and since the branches L_0^c and L_i^c have even length, $\operatorname{col} L_0^c = \operatorname{col} L_i^c = \operatorname{col} C$. Accordingly,

$$\begin{aligned}\Omega_{2j+1} &\in S(\operatorname{fin}(C, L_0^c)) \\ &= M(\operatorname{fin}(C, L_0^c)) \cap R(L_0^c) \subset R(L_0^c) \subset R(L_i^c), \\ \Omega_{2j+1} &\in M(\operatorname{fin}(C, L_i^c)) \cap R(L_i^c) \\ &= S(\operatorname{fin}(C, L_i^c)) = S(\operatorname{fin}(B, L_i)).\end{aligned}$$

If for some $i = h$ the move Ω_{2j+1} is admissible at the position $\operatorname{fin}(B, L_{h-1})$ but turns out to be non-admissible at $\operatorname{fin}(B, L_h)$ we test to see if it is allowed at $\operatorname{fin}(B, \Lambda_h)$.

Since $L_{h-1} = L_h + \Lambda_h$, the branch Λ_h is not empty, else the positions $\operatorname{fin}(B, L_{h-1})$ and $\operatorname{fin}(B, L_h)$ would coincide. If $\Omega_{2j+1} \in S(\operatorname{fin}(B, \Lambda_h))$ we set for $i = 1, 2, \ldots, h-1$

$$\begin{aligned}L_i &:= L_i * \Omega_{2j+1}; \\ L_i^c &:= L_i^c * \Omega_{2j+1}; \\ \Lambda_h &:= \Lambda_h * \Omega_{2j+1};\end{aligned}$$

and we go to the next step in the construction of the pseudocritical branch L_0 (see Figure 29). If however the move Ω_{2j+1} is not admissible at $\operatorname{fin}(B, \Lambda_h)$ under the rules of the original game $\mathfrak{A}$ or of the test model game $\mathfrak{S}_T(B)$ we analyze the reasons for the non-extendability (Case 2).

If the move Ω_{2j+1} is admissible (and allowed) at all the positions $\operatorname{fin}(B, L_i)$ for $i = 1, 2, \ldots, n$ it is admissible from the position $\operatorname{fin}(C, \Lambda_{an+1})$ also. In fact, $L_n^c = \Lambda_{n+1}$ and $\operatorname{fin}(C, \Lambda_{n+1}) = \operatorname{fin}(C, L_n^c) = \operatorname{fin}(B, L_n)$, since $L_n = L * L_n^c = L + \Lambda_{n+1}$. If

$$\begin{aligned}\Omega_{2j+1} &\in \tilde{S}(\operatorname{fin}(C, \Lambda_{n+1})) \\ &= M(\operatorname{fin}(C, \Lambda_{n+1})) \cap R(\Lambda_{n+1}) \setminus S(B),\end{aligned}$$

we set

$$\begin{aligned}L_i &:= L_i * \Omega_{2j+1}; \\ L_i^c &:= L_i^c * \Omega_{2j+1}; \\ \Lambda_{n+1} &:= \Lambda_{n+1} * \Omega_{2j+1}\end{aligned}$$

for $i = 1, 2, \ldots, n$ and pass to the next step in the construction of the pseudocritical branch.

If, however, the move Ω_{2j+1} is admissible but not allowed at the position $\operatorname{fin}(C, \Lambda_{n+1})$, it belongs to the set $S(B)$ and coincides with some move Ψ_i

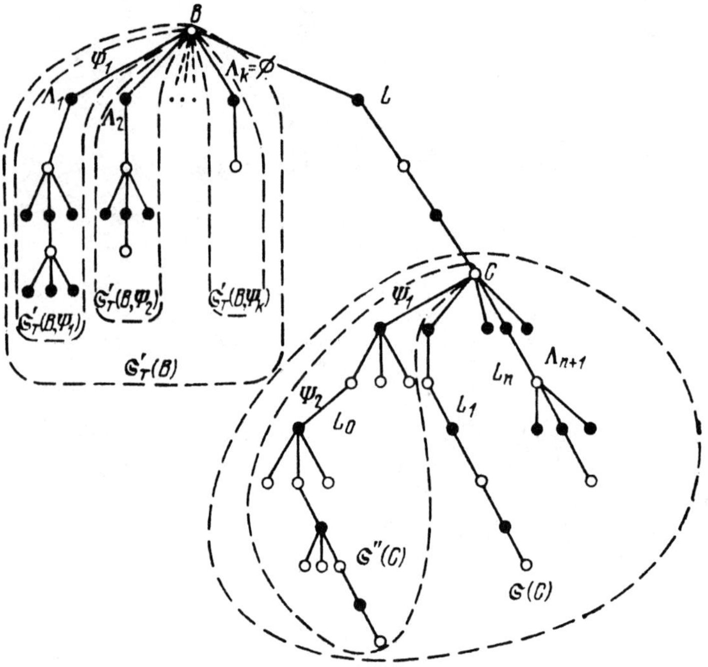

Figure 29

$(1 \le i \le n)$. Then it is allowed at the interior positions of the trees $\mathfrak{S}'_{\mathrm{T}}(B, \Psi_g)$ with the move belonging to $\operatorname{col} B = \operatorname{col} C$; these trees influence the tree $\mathfrak{S}_{\mathrm{T}}(B, \Psi_i)$, and the move is allowed at all positions of the latter where the turn to move belongs to the same color. (It is allowed at the analogous positions in the corresponding trees $\mathfrak{S}_{\mathrm{T}}(B, \Psi_g)$ and $\mathfrak{S}_{\mathrm{T}}(B, \Psi_i)$, but this has no significance for us.) It is clear that for this move to be allowed it must be admissible under the rules of the original game $\mathfrak{A}$.

Let Λ_h be a non-empty branch incident on the tree $\mathfrak{S}'_{\mathrm{T}}(B, \Psi_h)$ which influences the tree $\mathfrak{S}_{\mathrm{T}}(B, \Psi_i)$ $(1 \le h \le n)$, and let the move $\Omega_{2j+1} = \Psi_i$ be admissible under the rules of $\mathfrak{A}$ at the position $\operatorname{fin}(B, \Lambda_h)$. Since $L_{h-1} = L_h + \Lambda_h$, $L_{h-1} = L * L^c_{h-1}$ and $L_h = L * L^c_h$, the condition $L^c_{h-1} = L^c_h + \Lambda_h$ is satisfied. So $R(L^c_{h-1}) \subset R(\Lambda_h)$. The move $\Psi_i = \Omega_{2j+1}$ is allowed in the game $\mathfrak{S}(C)$ at the position $\operatorname{fin}(B, L_{h-1}) = \operatorname{fin}(C, L^c_{h-1})$. Accordingly

$$\begin{aligned}\Psi_i = \Omega_{2j+1} &\in S(\operatorname{fin}(C, L^c_{h-1})) \\ &= M(\operatorname{fin}(C, L^c_{h-1})) \cap R(L^c_{h-1}) \subset R(L^c_{h-1}) \subset R(\Lambda_h)\end{aligned}$$

and

$$\Psi_i = \Omega_{2j+1} \in M(\operatorname{fin}(B, \Lambda_h)) \cap R(\Lambda_h) \cap S(B).$$

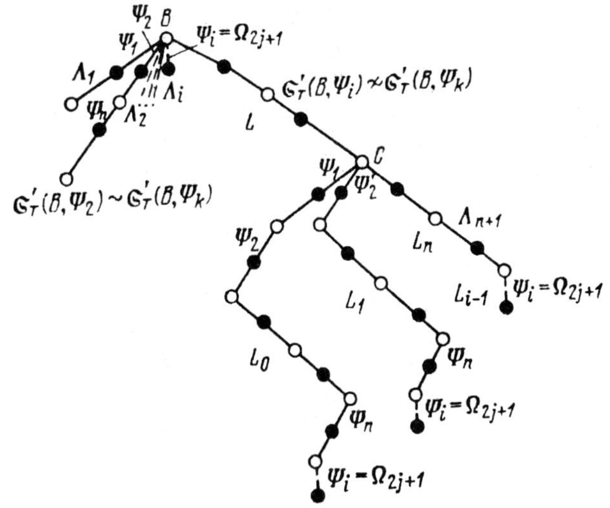

Figure 30

Moreover $\mathfrak{S}'_{\mathrm{T}}(B, \Psi_h) \sim \mathfrak{S}'_{\mathrm{T}}(B, \Psi_i)$. So, because of condition 3 of the theorem

$$\Psi_i = \Omega_{2j+1} \in S\big(\operatorname{fin}(B, \Lambda_h)\big).$$

We add this move at the end of a branch Λ_h such that the number $h \le n$ is maximal, and at the end of the branches $L_0, L_1, \ldots, L_{h-1}$; then we go on to the next step in constructing the pseudocritical branch L_0.

If the move $\Omega_{2j+1} = \Psi_{h'}$ is not admissible under the rules of the game $\mathfrak{A}$ at the position $\operatorname{fin}(B, \Lambda_n)$ we analyze the reasons for the non-extendability (case 3). Finally, if only empty branches Λ_i are incident on the trees $\mathfrak{S}'_{\mathrm{T}}(B, \Psi_i)$ that influence the tree $\mathfrak{S}'_{\mathrm{T}}(B, \Psi_{h'})$, e. g. for $j = 0$, we extend the branch $\Lambda_{h'}$ for $i = 0, 1, \ldots, h-1$:

$$L_i := L_i * \Omega_{2j+1}$$

$$L_i^c := L_i^c * \Omega_{2j+1}$$

$$\Lambda_{h'} := \Lambda_{h'} * \Omega_{2j+1} = \varnothing * \Omega_{2j+1} = \Lambda\big(\Omega_{2j+1}\big) = \Lambda(\Psi_{h'});$$

and we go to the next step (see Figure 30).

After we add the move Ω_{2j+1} of color col B to the pseudocritical branch L_0 all our conditions will be satisfied. In fact, let this move be added also to the branch Λ_h $(1 \le h \le n+1)$. By their construction, the branches Λ_i are not changed for $h \ne i$ and the move Ω_{2j+1} added at the end of Λ_h belongs to the corresponding tree $\mathfrak{S}'_{\mathrm{T}}(B, \Psi_h)$ for $1 \le h \le n$ or to $\tilde{\mathfrak{S}}'(C)$ for $h = n+1$.

Therefore the conditions $\Lambda_i \diamondsuit \mathfrak{S}'_T(B, \Psi_i)$ will be satisfied for $i = 0, 1, 2, \ldots, n$ and the condition $\Lambda_{n+1} \diamondsuit \mathfrak{S}'(C)$ will also be satisfied. The equations $L_{i-1} = L_i + \Lambda_j$, $L_i = L * L_i^c$, $L_i^c = \Lambda_{i+1} + \ldots + \Lambda_{n+1}$ will hold for $i = 0, 1, 2, \ldots, n$ since nothing changes on either side of the equations for $h \geq i$; for $h = i$ the move Ω_{2j+1} is added to the branches L_{i-1} and Λ_i, and for $h < i$ it is added to the branches L_{i-1}, L_{i-1}^c, L_i, L_i^c. The position $\text{fin}(B, L_i)$ belongs to $\mathfrak{S}(C)$ for $i = 1, 2, \ldots, n$ because it is allowed under the rules of $\mathfrak{S}(C)$ at the ends of the branches L_i, where the move Ω_{2j+1} is added. For a like reason $\text{fin}(B, L_0) \in \mathfrak{S}''(C)$ and $\text{fin}(B, L_n) \in \tilde{\mathfrak{S}}'(C)$.

If the new move Ω_{2j+1} is added at the end of a non-empty branch Λ_h, the non-empty branches Λ_i $(i \leq n)$ remain incident on the mutually non-influencing trees $\mathfrak{S}'_T(B, \Psi_i)$. Then they do not influence each other, and the empty branches Λ_i influence no branch whatsoever. The new non-empty tree $\Lambda_h(\Psi_h)$ is added only when the tree $\mathfrak{S}'_T(B, \Psi_h)$ influences no tree $\mathfrak{S}'_T(B, \Psi_i)$ on which the non-empty branch Λ_i is incident. Therefore the conditions connected with the mutual influences of the branches Λ_i and the trees $\mathfrak{S}'_T(B, \Psi_i)$ continue to hold in this case also.

Finally, the move Ω_{2j+1} was added at the end of some branch Λ_h $(1 \leq h \leq n+1)$ and at the ends of the branches $L_0, L_1, \ldots, L_{h-1}$, $L_0^c, L_1^c, \ldots, L_{h-1}^c$. Their lengths were odd, and the lengths of the branches Λ_i for $i \neq h$, and of L_i and L_i^c for $i \geq h$ remain even.

Now suppose that we have constructed the pseudocritical branch L_0 of odd length $\rho(L_0) = \rho(L) + 2j - 1$, where $j > 0$, that it satisfies all our conditions, and that the last move Ω_{2j-1}, of color $\text{col}\, C$, has been added at the ends of the branches Λ_h $(1 \leq h \leq n+1)$ and of the branches L_i for $i = 0, 1, \ldots, h-1$. If the position $\text{fin}(B, \Lambda_h)$ is terminal in the corresponding game $\mathfrak{S}_T(B)$ for $h \leq n$ or in $\tilde{\mathfrak{S}}(C)$ for $h = n+1$ we end the construction of L_0 and analyze the reasons for its non-extendability (Case 4). Otherwise we add the unique move Ω_{2j}, which belongs to either $S(\text{fin}(B, \Lambda_h))$ or $\tilde{S}(\text{fin}(C, \Lambda_{n+1}))$, to the corresponding minimal pruned tree $\mathfrak{S}'_T(B)$ or $\tilde{\mathfrak{S}}'(C)$ at the end of the branch Λ_h, of color opposite to $\text{col}\, B$. This move loses for $\text{col}\, B$ in the corresponding model game.

Suppose that $\Omega_{2j} \in M(\text{fin}(B, L_i)) = M(\text{fin}(C, L_i^c))$ for $i < h$. Since

$$\begin{aligned} L_i^c &= \Lambda_{i+1} + \ldots + \Lambda_h + \ldots + \Lambda_{n+1} \\ &= \Lambda_h + \tilde{\Lambda}_{i,h}, \end{aligned}$$

$$\Omega_{2j} \in M\big(\text{fin}\big(C, \Lambda_h + \tilde{\Lambda}_{i,h}\big)\big)$$

$$\cap\, S\big(\text{fin}(B, \Lambda_h)\big) \subset S\big(\text{fin}\big(C, \Lambda_h + \tilde{\Lambda}_{i,h}\big)\big) = S\big(\text{fin}\big(C, L_i^c\big)\big),$$

i.e. the move Ω_{2j} is allowed in the game $\mathfrak{S}(C)$. For $i = 0$ it leads from the position $\text{fin}(B, L_0)) \in \mathfrak{S}''(C)$ to a position in the tree $\mathfrak{S}''(C)$, which is a $\text{col}\, C$-pruned tree that also contains all the allowed moves at its positions of color opposed to $\text{col}\, C$.

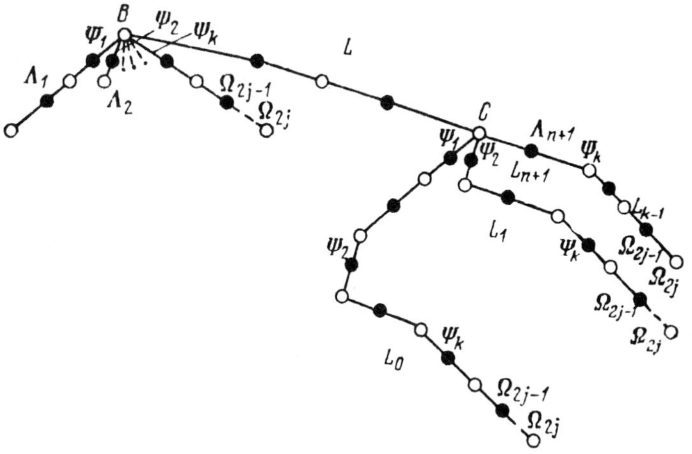

Figure 31

If the move Ω_{2j} is admissible, and therefore allowed, at all the positions $\operatorname{fin}(B, L_i)$ for $i < h$ we set (see Figure 31)

$$L_i := L_i * \Omega_{2j};$$

$$L_i^c := L_i^c * \Omega_{2j} \quad (i = 0, 1, \ldots, h-1)$$

$$\Lambda_h := \Lambda_h * \Omega_{2j}.$$

In the opposite case we end the construction of the pseudocritical branch L_0 and analyze the reasons for its non-extendability. (Case 5).

The branches L_i, L_i^c $(i = 0, 1, \ldots, h-1)$ and Λ_h will have even length after the move Ω_{2j} has been appended at their ends. For $i \geq h$ the lengths of L_i and L_i^c are unchanged and remain even.

The move Ω_{2j} belongs to the subtrees $\mathfrak{S}(C)$ and $\mathfrak{S}''(C)$, and to the corresponding subtree $\mathfrak{S}'_{\mathrm{T}}(B, \Psi_n) = \mathfrak{S}'_{\mathrm{T}}(B) \cap \mathfrak{S}_{\mathrm{T}}(B, \Psi_h)$ for $h \leq n$ or to $\tilde{\mathfrak{S}}'(C)$ for $h = n+1$. For $i < n$ it is added to both sides of the equations $L_{i-1} = L_i + \Lambda_i$, $L_i = L_i * L_i^c$, $L_i^c = \Lambda_{i+1} + \ldots + \Lambda_{n+1}$, and for $i \geq h$ the equations are untouched. This proves that our conditions are satisfied after we append to the pseudocritical branch L_0 the move Ω_{2j} of color opposite to col B just as they are when we append a move of color col B.

Thus we can proceed to the next step in the construction of L_0.

Let us go to the second and third stages of the proof. We consider, in turn, two cases in which L_0 is not extended.

(1) The pseudocritical branch L_0 has even length and its end $\operatorname{fin}(B, L_0) = \operatorname{fin}(C, L_0^c)$ is terminal in the game $\mathfrak{S}(C)$.

Since $S(B) \subset S(C)$ is a non-empty set the position C is non-terminal and some branch Λ_h contains at least two successive moves Ω_{2j-1} and Ω_{2j} of

L_0. It contains these and all the ends $\operatorname{fin}(B, L_i) = \operatorname{fin}(C, L_i^c)$ of the branches L_i and L_i^c for $i < h$. The position $\operatorname{fin}(B, L_0)$ is not lost in the game $\mathfrak{S}(C)$ and the move at it belongs to col C. Therefore it is terminal in the original game $\mathfrak{A}$ (new terminal positions in model games are lost for their own color).

We consider sequentially the positions $\operatorname{fin}(B, L_i)$ for $i = 0, 1, 2, \ldots, h-1$.

If the position $\operatorname{fin}(B, L_{i-1})$ is terminal and $\operatorname{fin}(B, L_i)$ is not, then since $L_{i-1} = L_i + \Lambda_i$, the axiom on Cases 3 and 4 implies that $\Lambda_i \sim L_i$.

If $\operatorname{fin}(B, L_{i-1})$ and $\operatorname{fin}(B, L_i)$ are both terminal, the axiom on Case 5 implies that either $\Lambda_i \sim L_i$ or $\operatorname{sc}(\operatorname{fin}(B, L_{i-1})) \underset{\text{col } B}{\preccurlyeq} \operatorname{sc}(\operatorname{fin}(B, L_i^i))$.

Thus if the branches Λ_i do not influence L_i for $i = 1, 2, \ldots, h-1$, all the positions $\operatorname{fin}(B, L_i)$ are terminal and

$$\operatorname{sc}(\operatorname{fin}(B, L_0)) \underset{\text{col } B}{\preccurlyeq} \operatorname{sc}(\operatorname{fin}(B, L_1)) \underset{\text{col } B}{\preccurlyeq} \cdots \underset{\text{col } B}{\preccurlyeq} \operatorname{sc}(\operatorname{fin}(B, L_{h-1})),$$

i.e. the position $\operatorname{fin}(B, L_{h-1})$ is also not lost. Since $\operatorname{fin}(B, L_n)$ belongs to $\mathfrak{S}'(C)$ it cannot be terminal and not lost; therefore $h - 1 < n$. Now let us look at the position $\operatorname{fin}(B, \Lambda_h) \in \mathfrak{S}'_{\mathrm{T}}(B, \Psi_h)$. It too cannot be terminal and not lost. Moreover, $L_{h-1} = L_h + \Lambda_h$. Hence the axiom on Cases 3 and 4, or the axiom on Case 5, implies that $\Lambda_h \sim L_h$. Thus we have shown that in the case we are considering there exist the branches Λ_i and L_i $(1 \le i \le n)$ that influence one another.

(2) The move Ω_{2j+1} of color col B is admissible at the position $\operatorname{fin}(B, L_{i-1})$ under the rules of the original game $\mathfrak{A}$ and, as we showed during the construction of the pseudocritical branch L_0, it is allowed in the game $\mathfrak{S}(C)$; it is admissible at $\operatorname{fin}(B, L_i)$ and at $\operatorname{fin}(B, \Lambda_i)$ it is either inadmissible or not allowed in the test model game $\mathfrak{S}_{\mathrm{T}}(B)$ for $1 \le i \le n$.

If it is inadmissible at $\operatorname{fin}(B, \Lambda_i)$ the conditions in the first axiom on Case 2 are satisfied, since $L_{i-1} = L_i + \Lambda_i$. Therefore $\Lambda_i \sim L_I$.

Suppose it is admissible but not allowed in the game $\mathfrak{S}_{\mathrm{T}}(B)$, i.e.

$$\Omega_{2j+1} \in M(\operatorname{fin}(B, \Lambda_i)) \setminus S(\operatorname{fin}(B, \Lambda_i)).$$

Since $L_{i-1} = L_i + \Lambda_i$, $L_{i-1} = L * L_{i-1}^c$ and $L_i = L * + L_i^c$, the condition $L_{i-1}^c = L_i^c + \Lambda_i$. The branch L_{i-1}^c has even length, and therefore col $L_{i-1}^c =$ col C. It consists of moves allowed by the rules of $\mathfrak{S}(C)$ at the corresponding positions. So, $L_{i-1}^c \diamondsuit \mathfrak{S}(C)$. At its end $\operatorname{fin}(C, L_{i-1}^c)$ the move Ω_{2j+1} is allowed, i.e. $\Omega_{2j+1} \in S(\operatorname{fin}(C, L_{i-1}^c))$. The branch Λ_i also has even length, and at its beginning B the move belongs to col $B =$ col C. Then by item 2 in the conditions of the theorem

$$\Omega_{2j+1} \in S(\operatorname{fin}(C, L_{i-1}^c)) = M(\operatorname{fin}(C, L_{i-1}^c)) \cap R(L_{i-1}^c) \subset R(L_{i-1}^c) \subset R(\Lambda_i).$$

Thus

$$\Omega_{2j+1} \in (M(\text{fin}(B, \Lambda_i)) \setminus S(\text{fin}(B, \Lambda_i))) \cap R(\Lambda_i).$$

According to item 3 of the conditions of the theorem

$$(M(\text{fin}\, B, \Lambda_i)) \setminus M(B)) \cap R(\Lambda_i) \subset S(\text{fin}(B, \Lambda_i)).$$

Therefore

$$\begin{aligned}\Omega_{2j+1} &\in (M(\text{fin}(B, \Lambda_i)) \setminus S(\text{fin}(B, \Lambda_i))) \cap R(\Lambda_i) \\ &= (M(\text{fin}(B, \Lambda_i)) \cap R(\Lambda_i)) \setminus S(\text{fin}(B, \Lambda_i)) \\ &\subset (M(\text{fin}(B, \Lambda_i)) \cap R(\Lambda_i)) \setminus ((M(\text{fin}(B, \Lambda_i)) \setminus M(B)) \\ &\quad \cap R(\Lambda_i)) = (M(\text{fin}(B, \Lambda_i)) \\ &\quad \setminus (M(\text{fin}(B, \Lambda_i)) \setminus M(B))) \cap R(\Lambda_i) = M(\text{fin}(B, \Lambda_i)) \\ &\quad \cap M(B) \cap R(\Lambda_i) \subset M(B).\end{aligned}$$

Thus

$$\begin{aligned}&\Omega_{2j+1} \in M(\text{fin}(B, L_{i-1}^c)) \cap M(\text{fin}(B, \Lambda_i)) \cap M(B), \\ &\Omega_{2j+1} \notin M(\text{fin}(B, L_i^c)), \\ &\quad L_{i-1}^c = L_i^c + \Lambda_i,\end{aligned}$$

i. e. the conditions for the second axiom on Case 2 are satisfied. Then either $\Lambda_i \sim L_i$ or $\Lambda_i \sim \Omega_{2j+1}$. But in the latter alternative the move Ω_{2j+1} would be allowed under the rules of the test model game $\mathfrak{S}_{\mathrm{T}}(B)$ at the position $\text{fin}(B, \Lambda_i)$ (by item 3 of the conditions of the theorem). Therefore the branch Λ_i influences the branch L_i.

(3) The move Ω_{2j+1} of color col B is admissible under the rules of $\mathfrak{A}$ at the positions $\text{fin}(B, L_{h-1})$ and $\text{fin}(B, L_h)$, and belongs to the set $S(B)$, so that it is admissible at B but not at $\text{fin}(B, \Lambda_h)$. $\Omega_{2j+1} = \Psi_{h'}$ for $1 \leq h' \leq n$ and if $h < i \leq n$ and the branch Λ_i is not empty, then $\mathfrak{S}'_{\mathrm{T}}(B, \Psi_i) \nsim \mathfrak{S}'_{\mathrm{T}}(B, \Psi_h)$.

Since $L_{h-1} = L_h + \Lambda_h$, the conditions for the second axiom on Case 2 are fulfilled in the present case. Thus either $L_h \sim \Lambda_h$ or $L_h \sim \Omega_{2j+1} = \Psi_{h'}$. If $L_h \sim \Lambda_h$, the axiom on the symmetry of the influence relation implies that $\Lambda_h \sim L_h$ (from now on we shall not mention this axiom explicitly, and will write $L'' \sim L'$ quite freely in place of $L' \sim L''$). But if $L_h \sim \Psi_{h'}$ we consider the non-empty branch $\tilde{\Lambda}_{h'}(\Psi_{h'}, \ldots\ )$, which is incident on the tree $\mathfrak{S}'_{\mathrm{T}}(B, \Psi_{h'})$, and has $\Psi_{h'} = \Omega_{2j+1}$ as its first move and necessarily exists, for example the branch $\Lambda(\Psi_{h'})$ consisting of the single move $\Psi_{h'}$. If for some i $(h < i \leq n)$ the branch Λ_i were to influence the move $\Psi_{h'}$ (it therefore could not be empty), the axiom of influence on a move and a branch would imply that it influenced the branch $\tilde{\Lambda}_{h'}$. Then by the definition of the influence relation of one tree on another, $\mathfrak{S}'_{\mathrm{T}}(B, \Psi_i)$ would influence $\mathfrak{S}'_{\mathrm{T}}(B, \Psi_{h'})$.

Therefore

$$\Lambda_i \nsim \Psi_{h'} \,|\, h < i \leq n.$$

(4) The end of a branch Λ_h $(1 \leq h \leq n+1)$ having odd length is a terminal position in the test model game $\mathfrak{S}_{\mathrm{T}}(B)$.

Since $\operatorname{fin}(B, \Lambda_h) \in \mathfrak{S}'_{\mathrm{T}}(B)$ and $\operatorname{col} \Lambda_h \neq \operatorname{col} B = \operatorname{col} C$, this position is terminal in the original game $\mathfrak{A}$ and, moreover, lost (as in the first situation where the pseudocritical branch L_0 could not be extended, a position which is terminal only in the model game must be lost for the color having the move there, i.e. the color leading to its branch—and $\operatorname{col} \Lambda_h \neq \operatorname{col} C$). Let us consider the positions $\operatorname{fin}(B, L_i)$ for $i = 0, 1, \ldots, h-1$. If there is a non-terminal position among them we choose the position $\operatorname{fin}(B, L_i)$ with the largest index $i < h$ for which $M(\operatorname{fin}(B, L_i)) \neq \varnothing$. Then $L_i = L_{i+1+\Lambda i+1}$ and $M(\operatorname{fin}(B, \Lambda_{i+1})) = \varnothing$, or if $i = h-1$, $M(\operatorname{fin}(B, \Lambda_{i+1})) = \varnothing$. Then by the axiom on Cases 3 and 4, $\Lambda_{i+1} \sim L_{i+1}$.

If, however, $M(\operatorname{fin}(B, L_i)) \neq \varnothing$ for $i = 1, 2, \ldots, h-1$, some of these positions are not lost; at least the position $\operatorname{fin}(B, L_0) \in \mathfrak{S}''(C)$ is not. We choose a non-lost terminal position $\operatorname{fin}(B, L_i)$ with maximal index $i < h$. Then either $\operatorname{fin}(B, L_{i+1})$ or (if $i = h-1$) $\operatorname{fin}(B, \Lambda_{i+1})$ is terminal and lost, i.e. for $i < h-1$

$$\operatorname{sc}(\operatorname{fin}(B, L_i)) \not\preccurlyeq_{\operatorname{col} B} \operatorname{sc}(\operatorname{fin}(B, L_{i+1})),$$

and for $i = h-1$

$$\operatorname{sc}(\operatorname{fin}(B, L_i)) \not\preccurlyeq_{\operatorname{col} B} \operatorname{sc}(\operatorname{fin}(B, \Lambda_{i+1})).$$

However, since $L_i = L_{i+1} + \Lambda_{i+1}$ the axiom on Case 5 implies that either $\Lambda_{i+1} \sim L_{i+1}$ or for $i < h-1$

$$\operatorname{sc}(\operatorname{fin}(B, L_i)) \preccurlyeq_{\operatorname{col} B} \operatorname{sc}(\operatorname{fin}(B, L_{i+1})),$$

and for $i = h-1$

$$\operatorname{sc}(\operatorname{fin}(B, L_i)) \preccurlyeq_{\operatorname{col} B} \operatorname{sc}(\operatorname{fin}(B, \Lambda_{i+1})).$$

The second alternative is incompatible with the assumption that $\operatorname{col} B$ loses at $\operatorname{fin}(B, L_{i+1})$ or $\operatorname{fin}(B, \Lambda_{i+1})$ and with the absence of a loss at $\operatorname{fin}(B, L_i)$. Thus $\Lambda_{i+1} \sim L_{i+1}$.

(5) The move Ω_{2j}, of color opposite to $\operatorname{col} B$, is admissible at the position $\operatorname{fin}(B, \Lambda_h)$ $(1 \leq h \leq n+1)$ under the rules of the game $\mathfrak{A}$ and inadmissible at $\operatorname{fin}(B, L_i)$ for some $i < h$.

In this case $i < n$ since $\operatorname{fin}(B, L_n) = \operatorname{fin}(C, \Lambda_{n+1})$. Just as in the two preceding cases we choose the position $\operatorname{fin}(B, L_i)$ with the largest value of i for which the move Ω_{2j} is admissible. Since $L_i = L_{i+1} + \Lambda_{i+1}$ the conditions for the axiom on Case 1 are satisfied. Then $L_{i+1} * \Omega_{2j} \sim \Lambda_{i+1}$ for $i < h-1$, and $\Lambda_{i+1} * \Omega_{2j} \sim L_{i+1}$ for $i=h-1$. In the first case $L_{i+1} * \Omega_{2j} \diamondsuit \tilde{\mathfrak{S}}'(C)$ and in the second case $\Lambda_{i+1} * \Omega_{2j} \diamondsuit \mathfrak{S}'_{\mathrm{T}}(B, \Psi_{i+1})$.

Thus, when the construction of the pseudocritical path L_0 is completed one of the following three conditions will be satisfied:

(1) Some branch Λ_i influences the corresponding auxiliary branch L_i $(1 \le i \le n)$.
(2) An auxiliary branch L_i influences the move $\Psi_{h'} \in S(B)$ $(1 \le i \le n$, $1 \le h' \le n)$, while $L_k \nsim \Psi_{h'}$ *for* $k = i+1, \ldots, n$.
(3) At the ends of the branches $L_{i+1}, \ldots, L_{-1}$, Λ_h $(1 \le i < h \le n)$ we may append a move Ω_{2j} of color opposite to col B:

$$L_k := L_l * \Omega_{2j} \qquad (k = i+1, \ldots, h-1);$$

$$\Lambda_h := \Lambda_h * \Omega_{2j}.$$

The conditions

$$L_{k-1} = L_k - \Lambda_k \qquad (k = i+2, \ldots, n);$$

$$\Lambda_{i+1} \sim L_{i+1}.$$

are satisfied. Now we show that some branch $\Lambda_k \diamondsuit \mathfrak{S}'_{\mathrm{T}}(B)$ $(1 \le k \le n)$ influences the branch $L_n \diamondsuit \tilde{\mathfrak{S}}'(B)$. This will complete the proof of the second alternative of the theorem on the transfer of scores of groups of moves—the influence of the tree $\mathfrak{S}'_{\mathrm{T}}(B)$ on the tree $\tilde{\mathfrak{S}}'(B)$—and therewith complete the proof of the theorem.

Suppose that for $1 \le i \le j \le k$ the branches Λ_i and L_j that we have constructed influence one another, or the branch L_j influences the move Ψ_i $(1 \le i \le n,\ 1 \le j \le n)$, while $\Lambda_k \nsim \Psi_i$ for $k = j+1, \ldots, n$. We have just shown that for some values of i and j one of these conditions is satisfied (if the first condition, $i = j$). If $j = n$ we have $\mathfrak{S}'_{\mathrm{T}}(B, \Psi_i) \sim \tilde{\mathfrak{S}}'($ B), i.e. $\mathfrak{S}'_{\mathrm{T}}(B) \sim \tilde{\mathfrak{S}}'(B)$, which is what we were to prove. In fact, either $\Lambda_i \sim L_n$ or, when $L_n \sim \Psi_i$, the axiom on the connection between influence on a move and influence on a branch implies that the branch L_n influences an arbitrary branch $\tilde{\Lambda}_i$ of the non-empty tree $\mathfrak{S}'_{\mathrm{T}}(B, \Psi_i)$, since such a branch contains its first move Ψ_i.

If $i, j < n$ we have $L_j \sim \Psi_i$, and since $L_j = L_{j+1} + \Lambda_{j+1}$ the first axiom on weak composition of branches implies that either $L_{j+1} \sim \Psi_i$ or $\Lambda_{j+1} \sim \Psi_i$ or $\Lambda_{j+1} \sim L_{j+1}$. The case $\Lambda_{j+1} \sim \Psi_i$ is ruled out by the relations we have proved above, and the other two have the form we have already studied, but with a larger index on the auxiliary branch L_{j+1}. If however $\Lambda_i \sim L_j$, the second axiom on the weak composition of branches implies that

$$\Lambda_i \sim L_{j+1} \qquad \text{or} \qquad \Lambda_i \sim \Lambda_{j+1} \qquad \text{or} \qquad \Lambda_{j+1} \sim L_{j+1}.$$

Here the second possibility is ruled out since the several branches Λ_k do not influence one another. The two remaining possibilities have the form we have already studied, but again the index of the branch L_{j+1} increases. So, in the end, we prove one of the relations

$$\Lambda_k \sim L_n \qquad (i < k \leq n) \qquad \text{or} \qquad \Lambda_n \sim \Psi_i,$$

as required.

In the 5th section we shall discuss in detail the use of our theorem for shortening the search. Here we shall only offer a few remarks. Suppose that at the position B we are considering the test model $\mathfrak{S}_{\mathrm{T}}(B)$ and the position $C = \operatorname{fin}(B, L)$, where the branch L has even length. If $L \nrightarrow \mathfrak{S}'_{\mathrm{T}}(B)$, we may begin by inspecting the moves $\Theta \in M(C) \setminus S(B)$ from C. At positions C with the move belonging to $\operatorname{col} C = \operatorname{col} B$ we need not consider moves belonging to $S(B)$. We define some minimal pruned tree $\tilde{\mathfrak{S}}'(C)$ of color opposite to $\operatorname{col}(B)$. If (a) the score (in the game $\tilde{\mathfrak{S}}(C)$) of all moves from C that belong to the tree $\tilde{\mathfrak{S}}'(B) = L \cup \mathfrak{S}'(C)$ is not better than the score of B in the test model game $\mathfrak{S}_{\mathrm{T}}(B)$, and (b) $\mathfrak{S}'_{\mathrm{T}}(B) \nrightarrow \tilde{\mathfrak{S}}'(B)$, the moves $\Psi \in S(B)$ from the position C may be neglected: $\operatorname{sc}(C) \preccurlyeq \operatorname{sc}_{\mathfrak{S}\mathrm{T}(B)}(B)$. A single test model $\mathfrak{S}_{\mathrm{T}}(B)$ may serve for several positions C_i $(i = 1, 2, \ldots, l)$, where $C_i = \operatorname{fin}(B, L^{(i)})$ and $\operatorname{col} L^{(i)} = \operatorname{col} B$.

A Constructive Definition of the Influence Relation for Several Games

Suppose that for some game $\mathfrak{A}$ we are given (a) an equivalence relation among moves from different positions, (b) the predicates $i(L, \Psi)$ and $I(B, L_1, L_2)$ for the influence relationships of the branch L on the move Ψ and of the branch L_1 with origin B on the branch L_2 with the same origin, and (c) the predicates for the influence of L on the subtree $\mathfrak{S}$ and the influence of a subtree $\mathfrak{S}_1$ on a subtree $\mathfrak{S}_2$, expressed in terms of the above elementary predicates:

$$I(B, L, \mathfrak{S}) := \bigvee I(B, L, \Lambda) | \Lambda \Diamond \mathfrak{S};$$

$$I(B, \Sigma_1, \mathfrak{S}_2) := \bigvee I(B, L_1, L_2) | L_1 \Diamond \mathfrak{S}_1, L_2 \Diamond \mathfrak{S}_2;$$

where the position B is the the origin of the branch L and the root of the subtrees $\mathfrak{S}$, $\mathfrak{S}_1$, and $\mathfrak{S}_2$. Suppose further that a theorem like the one we have proved above holds, which allows us to apply the method of analogies in cases where some branch or subtree does does not influence another subtree. Then the same theorem will also hold for the influence relationships defined by the more widely embracing predicate $I'(B, L_1, L_2)$ for which

$$B \in \mathfrak{A} \& L_1 \Diamond \mathfrak{A}(B) \& L_2 \Diamond \mathfrak{A}(B)$$
$$\Rightarrow I(B, L_1, L_2) \to I'(B, L_1, L_2)$$

where $\mathfrak{A}(B)$ is the B-subtree of the tree $\mathfrak{A}$.

Thus the influence relationships for a game $\mathfrak{A}$ can be defined in various ways. They may be distinguished from one another by the composition and amount of the information about moves and positions in the branches L_1 and L_2 defining the predicate $I(B, L_1, L_2)$, by the complexity of the algorithm for computing it, by the formulation of the theorem on the possibility of applying the method of analogy, and by the concept of equivalence among moves to which the method relates.

For example, if the theorem on the transfer of position scores is to hold, the influence relation need not be symmetric, but if the theorem on the transfer of scores of groups of moves is to hold, the influence relation must be symmetrized (at least partially). The larger the body of information used to define the value of the influence predicate and the more complex the predicates and the methods for using them, then usually the more often will the value of the predicate reduce to 0. Thus we find that we may be able to use the method of analogies effectively in shortening the search.

On the other hand, increasing the volume of information and complicating its structure leads to a growth in the amount of time needed per step in the search of the test model tree, since at each step we must evaluate the predicate. Moreover, the amount of memory needed will increase, since the computed information must be stored for future use.

Complicating the algorithms for evaluating the predicate and the methods for arriving at a decision by analogy—e. g. going over to methods using the notion of resolving a variation into what one might call independent components—leads to a growth in the time required for each step in the search since one must apply these algorithms and arrive at decisions.

We shall consider examples of a constructive definition of the influence relation, for the games of chess and noughts-and-crosses. We shall present predicates of several types, each having its own advantages and weaknesses. To simplify the reasoning we shall substitute for chess a nearly equivalent game that we have mentioned earlier. In it one need not defend against checks, the King may attack, and the game is won by capturing the opponent's King. If stalemate is not counted as a draw, but as a loss for the stalemated side, this game will be equivalent to chess.

Our model allows all moves admissible under the rules of chess, and some others as well. The latter lead to the inclusion of auxiliary positions in the tree of the game itself and of the models used in game programming, and lead also to the arising of 'parasitic' influences from moves that either do not defend one's King from a check, or put him under attack. This effect is clearly more important than the non-equivalence of our game to chess. The latter, in general, is irrelevant when—as is usually the case—we are contemplating a model with limited depth of search, and mate or stalemate are far away.

We shall present (but without supporting it) an influence predicate for chess which takes into account checks to the King and the pinning of pieces that protect it from attacks by long-range pieces. We have mentioned earlier the proofs of theorems on the transfer of scores of positions and groups of moves [15]; these were carried out for the influence relation defined in them. We again assume that there are neither castlings nor captures *en passant* among the moves in the branches we consider. The rules of castling preclude its execution under attacks by enemy pieces on the squares where the King stands before and after castling or on the intermediate squares. We neglect such restrictions, as connected with attacks on the King. A capture *en passant* is the sole chess move in which the capturing piece does not occupy the square on which the captured piece stood. We could to some extent alter the influence predicate to take account of these captures, but it is simpler to treat the branches containing them as influencing arbitrary ones.

To construct concrete influence predicates we study the following scheme. Suppose given the branches $L_1(\Psi_1, \Psi_2, \ldots, \Psi_k)$ and $L_2(\Theta_1, \Theta_2, \ldots, \Theta_l)$ together with their segments $L_1^g(\Psi_1, \Psi_2, \ldots, \Psi_g)$ and $L_2^h(\Theta_1, \Theta_2, \ldots, \Theta_h)$ $(1 \leq g \leq k,\ 1 \leq h \leq l)$. We consider the composition $L^{g,h} = L_1^g + L_2^h$ of these segments. An influence relation will exist when one of the cases of Type 1 or Type 2 arises:

at the position $\mathrm{fin}(B, L^{2\gamma-1,2\eta})$ the move $\Psi_{2\gamma}$ of color opposite to col B is inadmissible;

at the position $\mathrm{fin}(B, L^{2\gamma,2\eta-1})$ the move $\Theta_{2\eta}$ of the same color is inadmissible; or

at the position $\mathrm{fin}(B, L^{2\gamma,2\eta})$ it is possible to make a move Ω of color col B that is inadmissible at the positions $\mathrm{fin}(B, L_1^{2\gamma})$ and $\mathrm{fin}(B, L_2^{2\eta})$, but in certain circumstances at only one of these (see Figure 32). Information on the branches and subtrees investigated during the search will consist of some set of boards. The definition of a board was given in the preceding chapter, but we shall recall it here. A board is an array of 64 bits, corresponding to the squares on a chessboard; the bits corresponding to squares that are in some way distinguished are set to 1 and the remainder to

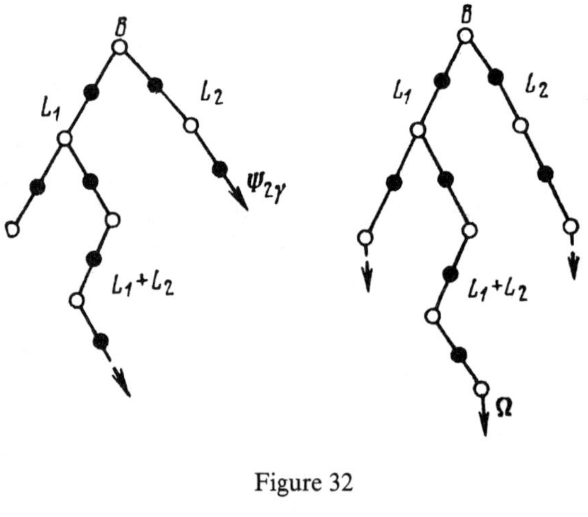

Figure 32

0. The union $T_1 \cup T_2$ of two boards is a board on which a bit is set to 1 if and only if it is set to 1 on one or both of T_1 and T_2; the intersection $T_1 \cap T_1$ is a board in which a bit is set to 1 if and only if it is set to 1 on both T_1 and T_2; in the non-coincidence, or discrepancy board $T_1 \oplus T_2$ a bit is set to 1 if and only if its values differ on the two boards; in the difference board $T_1 \setminus T_2$when it has the value 1 on T_1 and 0 on T_2; in the inverse board $\neg T$ a bit is set to 1 if it is 0 in T, and to 0 if 1 in T. The predicate $[T]$ has the value 'true' if T is not empty, i. e. if at least one bit in T is set to 1, and 'false' if $T = \varnothing$, i. e. all bits in T are set to 0.

We introduce the following information elements that describe the virtual chess move Ψ:

(1) $E(\Psi)$—a board on which the square where the move originates is distinguished (bit set to 1);
(2) $I(\Psi)$—the board for the square to which the piece moved;
(3) $W(\Psi)$—a board containing the trajectory of the move.

For moves of long-range pieces—Bishops, Rooks, and Queens—the board $W(\Psi)$ contains marks on the squares in diagonals, ranks, and files as appropriate for the move between the squares $E(\Psi)$ and $I(\Psi)$. It is empty for other moves, $W(\Psi) = \varnothing$.

Moves of the color col B will be considered as ours; moves of the opposite color will be moves by the opponent. The boards for these and the other moves in a contemplated branch or subtree will be compounded. Thus

$$E(s, \mathfrak{L}) := \bigcup E(\Psi) | \Psi \in \Lambda, \operatorname{col} \Psi = \operatorname{col} B;$$

$$E(c, \mathfrak{L}) := \bigcup E(\Psi) | \Psi \in \Lambda, \operatorname{col} \Psi \neq \operatorname{col} B;$$

$$I(s, \mathfrak{L}) := \bigcup I(\Psi) | \Psi \in \Lambda, \operatorname{col} \Psi = \operatorname{col} B;$$

$$I(c, \mathfrak{L}) := \bigcup I(\Psi) | \Psi \in \Lambda, \operatorname{col} \Psi \neq \operatorname{col} B;$$

$$W(c, \mathfrak{L}) := \bigcup W(\Psi) | \Psi \in \Lambda, \operatorname{col} \Psi \neq \operatorname{col} B;$$

where $\mathfrak{L}$ is the contemplated branch or subtree. We do not need the board $W(\mathrm{s}, \mathfrak{L})$.

Let $P(A, \mu)$ be the board for the *mobility* of the piece μ in the position A. This board distinguishes those squares that are attacked by μ from A. If such a square is not occupied by a piece of the same color, the piece μ can move to it. Pawns are an exception, since they do not move in the same way as they attack. For this reason there are two boards corresponding to a Pawn μ: $p(A, \mu)$ for the squares to which it can move, and $b(A, \mu)$ for the squares it attacks. We denote by $\lambda(A; \Psi_1, \Psi_2)$ the branch beginning at A and consisting of the two moves Ψ_1 and Ψ_2, naturally of opposite color. We shall from time to time contemplate a short branch $\lambda(A; \Psi)$ consisting of one move. The fact that such branches are contained in our contemplated branch L with origin B, or in the subtree $\mathfrak{S}$, is expressed, together with their color, by the following statements:

$$\lambda(A;\Psi_1,\Psi_2)\in L := L = L_1 * \lambda * L_2 \& A = \operatorname{fin}(B, L_1);$$
$$\lambda(A;\Psi)\in L := L = L_1 * \lambda \& A = \operatorname{fin}(B, L_1);$$
$$\lambda \in \mathfrak{S} := \exists L(L\Diamond \mathfrak{S} \& \lambda \in L);$$
$$\operatorname{col}\lambda(A;\Psi_1,\Psi_2) := \operatorname{col}\lambda(A;\Psi) := \operatorname{col} A.$$

In the first of these definitions we do not require that the branch L_2 consist of admissible moves when its origin is the position B. It must be admissible only at the position $\operatorname{fin}(B, L_1 * \lambda)$.

We shall need the so-called 'boards of new mobility',

$$N(\lambda,\mu) := \begin{cases} P(\operatorname{fin}(A,\lambda),\mu), & \text{if } \mu = \varphi(\Psi'): \\ P(\operatorname{fin}(A,\lambda),\mu)\setminus P(A,\mu), & \text{if } \mu \neq \varphi(\Psi'), \end{cases} \quad \mu\text{—not a pawn}$$

$$n_{p,b}(\lambda,\mu) := \begin{cases} p, b(\operatorname{fin}(A,\lambda),\mu), & \text{if } \mu = \varphi(\Psi'); \\ \varnothing, & \text{if } \mu \neq \varphi(\Psi'), \end{cases} \quad \mu\text{—a pawn}$$

where A is the origin of the branch and Ψ is its first and perhaps only move. We also compound these boards for all pieces and two-move branches of the same color in the contemplated branch or subtree, and include as well the single-move branches that are ends of branches L of odd length:

$$N(s,\mathfrak{L}) := \bigcup N(\lambda,\mu) | \lambda \in \mathfrak{L}, \mu \in \Phi, \operatorname{col}\lambda = \operatorname{col}\mu = \operatorname{col} B;$$
$$n_{p,b}(s,\Lambda) := \bigcup n_{p,b}(\lambda,\mu) | \lambda \in \mathfrak{L}, \mu \in \Pi, \operatorname{col}\lambda = \operatorname{col}\mu = \operatorname{col} B;$$

where Φ is a set of pieces (Kings, Queens, Rooks, Bishops, and Knights) and Π a set of Pawns. We do not need the analogous boards $N(c,\mathfrak{L})$ and $n_{p,b}(c,\mathfrak{L})$.

Let some enemy move $\varphi(\Psi_{2\gamma})$ in the branch L_1 be inadmissible at the position $\operatorname{fin}(B, L_1^{2\gamma-1} + L_2^{2\eta})$. This can happen only in the following cases:

(1) The enemy piece $\varphi(\Psi_{2\gamma})$ does not occupy the square $E(\Psi_{2\gamma})$ at the position $\operatorname{fin}(B, L_1^{2\gamma-1} + L_2^{2\eta})$. Since it occupies this square at the position $\operatorname{fin}(B, L_1^{2\gamma-1})$ it either left it by one of the moves $\Theta_{2\varepsilon} \in L_2$, of color opposite to col B, or it was captured by some move $\Theta_{2\delta+1} \in L_2$ of color col B ($2\varepsilon \leq 2\eta$ or $2\delta+1 < 2\eta$). Therefore

$$E(c,L_1)\cap(E(c,L_2)\cup I(s,L_2)) \neq \varnothing.$$

(2) There is no piece $\chi(\Psi_{2\mu})$ on the square $I(\Psi_{2\gamma})$. If $\chi(\Psi_{2\gamma}) \neq \varnothing$, i. e. if the move $\Psi_{2\gamma}$ is a capture, the piece $\chi(\Psi_{2\gamma})$ left this square by one of the moves $\Theta_{2\delta+1} \in L_2$ of color col B or it was captured by a move $\Theta_{1\varepsilon} \in L_2$ of the opposite color. If however $\chi(\Psi_{2\gamma}) = \varnothing$, the square $I(\Psi_{2\gamma})$ was occupied in one of the moves $\Theta \in L_2$ ($2\delta+1$ or 2ε or h cannot exceed 2η). Then

$$I(c,L_1)\cap(E(s,L_2)\cup I(s,L_2)\cup I(c,L_2)) \neq \varnothing.$$

(3) The trajectory of the move $W(\Psi_{2\gamma})$ is partitioned by some move $\Theta_h \in L_2$. Then

$$W(c, L_1) \cap (I(s, L_2) \cup I(c, L_2)) \neq \varnothing .$$

For similar reasons a move $\Psi_{2\eta}$ by the opponent may be inadmissible at the position $\text{fin}(B, L_1^{2\gamma} + L_2^{2\eta-1})$.

Let us now consider the cases in which at the position $\text{fin}(B, L_1^{2\gamma} + L_2^{2\eta})$ it is possible to make a move Ω (of color col B) which is inadmissible at both of the positions $\text{fin}(B, L_1^{2\gamma})$ and $\text{fin}(B, L_2^{2\gamma})$. Every piece in the position $\text{fin}(B, L_1^{2\gamma} + L_2^{2\eta})$ is on the same square that it occupies at the end of one or other or both of the branches $L_1^{2\gamma}$ and $L_2^{2\eta}$ issuing from B. Then the piece $\varphi(\Omega)$ is on the square $E(\Omega)$ in one or the other, or both, of the positions $\text{fin}(B, L_1^{2\gamma})$ and $\text{fin}(B, L_2^{2\gamma})$. In the first case, since $\varphi(\Omega)$ is on the same square $E(\Omega)$ in the position $\text{fin}(B, L_1^{2\gamma} + L_2^{2\eta})$ but not in the position $\text{fin}(B, L_2^{2\eta})$ it occupied this position in one of the moves $\Psi_{2\varepsilon+1} \in L_1$ $(\varepsilon < \gamma)$. Thus all the squares it can reach from the position $\text{fin}(B, L_1^{2\gamma})$ belong to one of the positions on the new possibility boards $N(\lambda, \varphi(\Omega))$, $n_p(\lambda, \varphi(\Omega))$ or $n_b(\lambda, \varphi(\Omega))$ (the latter cases hold when $\varphi(\Omega)$ is a Pawn) for one of the two-move branches $\lambda \in L_1^{2\gamma}$ having the color col B. The move Ω is of course not allowed from the position $\text{fin}(B, L_2^{2\eta})$ but may turn out to be so from the position $\text{fin}(B, L_2^{2\gamma})$ for the following reasons.

(1) The square $I(\Omega)$ is not occupied by the same piece that it holds in the position $\text{fin}(B, L_1^{2\gamma} + L_2^{2\eta})$. If $\chi(\Omega) \neq \varnothing$, the piece $\chi(\Omega)$ arrived at the given square in one of the moves $\Theta_{2\varepsilon} \in L_2 (\varepsilon \leq \eta)$, i.e. $I(\Omega) = I(\Theta_{2\varepsilon})$. If however $I(\Omega) = \varnothing$, a piece that occupies it in the position $\text{fin}(b, L_1^{2\gamma})$ or arrives on it in some move $\Theta_g \in L_2$, must have left it during one of the moves $\Theta_{2\varepsilon} \in L_2$ $(n \leq 2\eta)$. Accordingly,

$$I(\Omega) \in E(s, L_2) \cup E(c, L_2) \cup I(c, L_2)$$

and, since the square $I(\Omega)$ belongs to one of the boards $N(\lambda, \varphi(\Omega))$, $n_p(\lambda, \varphi(\Omega))$ or $n_b(\lambda, \varphi(\Omega))$ for some two-move branch $\lambda \in L_1^{2\gamma}$ with color col B, the following conditions are satisfied:

$$N(s, L_1) \cap (E(s, L_2) \cup E(c, L_2) \cup I(c, L_2)) \neq \varnothing$$

$$n_p(s, L_1) \cap E(s, L_2) \cup E(c, L_2) \neq \varnothing$$

$$n_b(s, L_1) \cap I(c, L_2) \neq \varnothing$$

(2) The piece $\chi(\Omega)$ occupies the square $I(\Omega)$ in the position $\text{fin}(B, L_1^{2\gamma})$, but the trajectory of the move Ω is interrupted by some other piece. Since this trajectory is free in the position $\text{fin}(B, L_1^{2\gamma} + L_1^{2\eta})$, all of the blocking pieces leave it during one or more moves $\Theta_{ih} \in L_2^{2\gamma}$ $(h \leq 2\gamma)$. The ex-blocking-piece nearest to the square $E(\Omega)$ will be found in the position $\text{fin}(B, L_2^{2\gamma})$ under attack by the long range piece $\varphi(\Omega)$, and the square it occupies belongs to the board $N(s, L_1)$. It may happen that it is not this piece that leaves this square, but rather some capturing piece

in the branch L_2, or a piece capturing the latter, and so on. In any case

$$N(s, L_1) \cap E(s, L_2) \cup E(c, L_2) \neq \varnothing$$

The case in which the piece $\varphi(\Omega)$ occupies the square $E(\Omega)$ in the position $\operatorname{fin}(B, L_2^{2\eta})$ but not in $\operatorname{fin}(B, L_1^{2\gamma})$ is symmetric with the case we have just considered. Therefore we have only the case to consider in which this piece occupies the same square $E(\Omega)$ in all three of the positions $\operatorname{fin}(B, L_2^{2\gamma})$, $\operatorname{fin}(B, L_2^{2\eta})$, and $\operatorname{fin}(B, L_1^{2\gamma} + L_2^{2\eta})$. Then it occupied this same square in the position B. Otherwise it would have arrived at this square in both of the branches $L_1^{2\gamma}$ and $L_2^{2\eta}$, and if it arrived there more than once the corresponding number of arrivals must have been greater by 1 than the number of departures. But then in the branch $L_1^{2\gamma} + L_2^{2\eta}$ the number of arrivals must have been greater by 2 than the number of departures, and this is impossible. Moreover, as we have already noted, in at least one of the positions $\operatorname{fin}(B, L_1^{2\gamma})$ or $\operatorname{fin}(B, L_2^{2\eta})$ the piece $\chi(\Omega)$ (perhaps a null piece) occupies the square $I(\Omega)$. Then two cases may exist.

(1) The move Ω is admissible at both of the positions $\operatorname{fin}(B, L_1^{2\gamma})$ and $\operatorname{fin}(B, L_2)$. This case is of no interest.

(2) The move Ω is admissible at one and only one of the positions $\operatorname{fin}(B, L_1^{2\gamma})$ and $\operatorname{fin}(B, L_2^{2\eta})$. If it is inadmissible at B, the case is of no interest. But it cannot be admissible at B, else the piece $\chi(\Omega)$ would there occupy the square $I(\Omega)$ and the trajectory $W(\Omega)$ would be either empty or free. Suppose that Ω is inadmissible at $\operatorname{fin}(B, L_2^{2\eta})$. Then our King is not taken (it would be taken also at $\operatorname{fin}(B, L_1^{2\gamma} + L_2^{2\eta})$). Thus either the piece $\chi(\Omega)$ is not on the square $I(\Omega)$ or the trajectory $W(\Omega)$ is blocked. In the latter case there was a piece on some square $p \in W(\Omega)$; we shall count the number of arrivals and departures of this piece with respect to the squares $I(\Omega)$ and p in the branches $L_1^{2\gamma}$, $L_2^{2\eta}$, and $L_1^{2\gamma} + L_2^{2\eta}$.

Since the move Ω is admissible at $\operatorname{fin}(B, L_1^{2\gamma})$ the number of exits of the piece $\chi(\Omega)$ at the square $I(\Omega)$ equals the number of its returns (the exit of a null piece $\chi(\Omega) = \varnothing$ will be regarded as an occupation and its arrival as a freeing of the square). In the same way, the number of arrivals of any piece at the square $p \in W(\Omega)$ must equal the number of its departures. At the position $\operatorname{fin}(B, L_1^{2\gamma})$ the move Ω is inadmissible; this means that the piece $\chi(\Omega)$ has made an uncompensated exit from the square $I(\Omega)$ or that some piece has made an uncompensated entrance at the square p. The number of arrivals at a square and exits from it in the branch $L_1^{2\gamma} + L_2^{2\eta}$ equals the sum of the corresponding numbers of arrivals and exits in the constituent branches $L_1^{2\gamma}$ and $L_2^{2\eta}$. Thus at the position $\operatorname{fin}(B, L_1^{2\gamma} + L_2^{2\eta})$ the piece $\chi(\Omega)$ is not on the square $I(\Omega)$, or there is a piece on the square p: the move Ω is inadmissible. We treat in the same way the case in which Ω is admissible only at the position $\operatorname{fin}(B, L_2^{2\eta})$. So, the conditions for applying the second axiom on Situation 2 are not satisfied.

(3) The move Ω is inadmissible at $\operatorname{fin}(B, L_1^{2\gamma})$ because the piece $\chi(\Omega)$ is not on the square $I(\Omega)$, and is inadmissible at $\operatorname{fin}(B, L_2^{2\eta})$ because the trajectory $W(\Omega)$ is blocked. It cannot have been free at B, else it would also be free at $\operatorname{fin}(B, L_2^{2\eta})$ or blocked at $\operatorname{fin}(B, L_1^{2\gamma})$ {a piece on this trajectory in the branch $L_2^{2\eta}$ cannot leave it in the branch $L_1^{2\gamma}$}. Thus in the position $\operatorname{fin}(B, L_1^{2\gamma})$ the square $I(\Omega)$ is attacked by the piece $\varphi(\Omega)$ as a result of the move $\Psi \in L_1$ $(h \leq 2\gamma)$, and $I(\Omega) \in N(s, L_1)$. On the other hand, the piece $\chi(\Omega)$ enters the square $I(\Omega)$ in the branch $L_2^{2\eta}$ (If $\chi(\Omega) = \varnothing$, the square $I(\Omega)$ becomes free in the branch $L_2^{2\eta}$.) Therefore

$$N(s, L_1) \cap (E(sL_2) \cup E(c, L_2) \cup I(c, L_2)) \neq \varnothing .$$

(4) The trajectory $W(\Omega)$ is freed of friendly pieces by moves in the branches L_1 and L_2. Among the squares thus freed, let the square that is nearest to $E(\Omega)$ be freed by a move in L_1, and among the squares freed by a move in L_2 let p be that one nearest to $E(\Omega)$. Then in the position B the square p is not under attack by the piece $\varphi(\Omega)$, but is under attack by it in the position $\operatorname{fin}(B, L_1^{2\gamma})$. Therefore $p \in N(s, L)$. At the same time,

$$p \in E(s, L_2) \cup E(c, L_2)) \neq \varnothing .$$

(5) The move Ω is the initial jump of a Pawn from the second rank to the fourth, or from the seventh to the fifth, while one of the squares blocking it is freed by a move in $L_1^{2\gamma}$ and another by a move in $L_2^{2\eta}$. Let hor_ν be a board on which the squares in the ν-th rank are marked, let $T^{\uparrow}$ represent a displacement of the squares on the board T by one rank upward, and $T^{\downarrow}$ by one rank downward. Let $\Pi(s, B)$ be a board on which our own Pawns stand in the position B. If White has the move at B

$$\begin{aligned} G(B, L_1, L_2) &:= \\ &= \Pi(s, B)^{\uparrow} \cap \operatorname{hor}_3 \cap \Big(\big((E(s, L_1) \cup E(c, L_1))^{\downarrow} \\ &\qquad \cap (E(s, L_2 \cup E(c, L_2))) \\ &\qquad \cup \big((E(s, L_1) \cup E(c, L_1)) \cap (E(s, L_2) \cup E(c, L_2))^{\downarrow} \big) \Big) \\ &\neq \varnothing ; \end{aligned}$$

If Black has the move,

$$\begin{aligned} G(B, L_1, L_2) &:= \\ &= \Pi(s, B)^{\downarrow} \cap \operatorname{hor}_6 \cap \Big(\big((E(s, L_1) \cup E(c, L_1))^{\uparrow} \\ &\qquad \cap (E(s, L_2) \cup E(c, L_2)) \big) \\ &\qquad \cup \big((E(s, L_1) \cup E(c, L_1)) \cap (E(s, L_2) \cup E(c, L_2))^{\uparrow} \big) \Big) \\ &\neq \varnothing . \end{aligned}$$

For some models this case need not be taken into account.

The influence predicate Inf$((B, L_1, L_2)$ takes on the value *True* when one of the above conditions, or its symmetric counterpart, is satisfied. Thus

$$\begin{aligned}
\mathrm{Inf}(B, L_1, L_2) &:= \\
&= \Big[(E(c,L_1)\cap E(c,L_2))\cup((I(c,L_1)\cup W(c,L_1)) \\
&\qquad \cap(I(s,L_2)\cup I(c,L_2)))\cup((I(s,L_1)\cup I(c,L_1)) \\
&\quad \cap(I(c,L_2)\cup W(c,L_2))) \\
&\quad \cup\big((I(c,L_1)\cup N(s,L_1)\cup n_p(s,L_1))\cap(E(s,L_2)\cup E(c,L_2))\big) \\
&\quad \cup\big((E(s,L_1)\cup E(c,L_1))\cap(I(c,L_2)\cup N(s,L_2)\cup n_p(s,L_2))\big) \\
&\quad \cup((N(s,L_1)\cup n_b(s,L_1))\cap I(c,L_2))\cup(I(c,L_1) \\
&\qquad \cap(N(s,L_2)\cup n_b(s,L_2)))\cup G(B,L_1,L_2)\Big].
\end{aligned}$$

This formula is more convenient for calculation, since during the traversal of branches or subtrees we need to form a limited number of boards, and in testing the influence of branches on subtrees or of subtrees on subtrees we need to form a strongly limited number of unions and intersections. However, to study the properties of the influence predicate it is better to use a different formulation of it, which we now proceed to define.

Every board on which it depends is the union of two well-defined boards corresponding to the two-move branches $\lambda(A;\Psi',\Psi'')$ belonging to the branches L under consideration, with origin B, and the single-move branches $\lambda'(A;\Psi)$ with origins at the ends of branches of odd length (the latter may be formally considered to be two-move branches, with a null second move and empty boards corresponding to it). All these short branches have the color col B. Since the operations of union and intersection are distributive, the formula given above for the influence predicate can be written in the form

$$\mathrm{Inf}(B, L_1, L_2) = \left[\bigcup_{\lambda_1\in L_1}\bigcup_{\lambda_2\in L_2}\bigcup_{\zeta\in Z}\left(T_{\sigma_1(\zeta)}(\lambda_1)\cap T_{\sigma_2(\zeta)}(\lambda_2)\right)\right]$$

where Z is the completely determined set of intersections of (a) the boards $E(\Psi)$, $I(\Psi)$, $W(\Psi)$, $N(\lambda,\mu)$, $n_p(\lambda_1,\mu)$ and $n_b(\lambda_1,\mu)$ corresponding either to the branch $\lambda_1\in L_1$ or to its moves Ψ (we denote these boards by the symbol $T_{\sigma_1(\zeta)}(\lambda_1)$) and (b) the analogous boards $T_{\sigma_2(\zeta)}(\lambda_2)$ corresponding to the branch $\lambda_2\in L_2$.

Let us dwell on the question of how we prove theorems on the transfer of scores of positions and the scores of groups of moves for the influence predicate we have just constructed. We do not need the axioms on the relationship between influence on moves and on a branch, nor the first axiom on the composition of branches, since we do not need to consider the influence of branches on moves. The axiom of symmetry is satisfied since the influence predicate $\mathrm{Inf}(B, L_1, L_2)$ is symmetric with respect to the boards corresponding to the branches L_1 and L_2. We use the second axiom on the composition of branches when the branch $\Lambda_1 + \Lambda_2$ is composed of two-move branches λ of color col B belonging entirely either to Λ_1 or to Λ_2 (with perhaps single-move branches λ' of color col B, lying at the end of one of the branches Λ_1, Λ_2 and at the end of the branch $\Lambda_1 + \Lambda_2$).

The influence predicate $\mathrm{Inf}(B, \Lambda_1 + \Lambda_2, \Lambda_3)$ can be written in the form

$$\begin{aligned}
&\mathrm{Inf}(B, \Lambda_1 + \Lambda_2, \Lambda_3)\\
&\quad = \left[\bigcup_{\lambda_1 \in \Lambda_1 + \Lambda_2} \bigcup_{\lambda_2 \in \Lambda_2} \bigcup_{\zeta \in Z} \left(T_{\sigma_1(\zeta)}(\lambda_1) \cap T_{\sigma_2(\zeta)}(\lambda_2) \right) \right]\\
&\quad = \left[\bigcup_{\lambda_1 \in \Lambda_1} \bigcup_{\lambda_2 \in \Lambda_2} \bigcup_{\zeta \in Z} \left(T_{\sigma_1(\zeta)}(\lambda_1) \cap T_{\sigma_2(\zeta)}(\lambda_2) \right) \right]\\
&\qquad \cup \left[\bigcup_{\lambda_1 \in \Lambda_2} \bigcup_{\lambda_2 \in \Lambda_3} \bigcup_{\zeta \in Z} \left(T_{\sigma_1(\zeta)}(\lambda_1) \cap T_{\sigma_2(\zeta)}(\lambda_2) \right) \right]\\
&\quad = \widetilde{\mathrm{Inf}}(B, \Lambda_1, \Lambda_3) \vee \widetilde{\mathrm{Inf}}(B, \Lambda_2, \Lambda_3),
\end{aligned}$$

where the predicates $\widetilde{\mathrm{Inf}}(B, \Lambda_1, \Lambda_3)$ and $\widetilde{\mathrm{Inf}}(B, \Lambda_2, \Lambda_3)$ differ from the predicates $\mathrm{Inf}(B, \Lambda_1, \Lambda_3)$ and $\mathrm{Inf}(B, \Lambda_2, \Lambda_3)$ only in the fact that the boards $T_\sigma(\lambda)$ corresponding to the two-move branches λ and belonging to the branches Λ_1 and Λ_2 are intersected in $\widetilde{\mathrm{Inf}}$ by the boards $T_\sigma(\lambda')$ corresponding to the branches λ' composed of the same moves but belonging to the composite branch $\Lambda_1 + \Lambda_2$. But, all of the boards $T_\sigma(\lambda)$, except $Nf(\lambda, \mu)$ for the long-range pieces (Queens, Rooks, Bishops), are defined by the moves Ψ' and Ψ'' in the branch λ and do not depend on its origin A. This means that for these boards $T_\sigma(\lambda) = T_\sigma(\lambda')$.

Let $\mathrm{Inf}(B, \Lambda_1 + \Lambda_2, \Lambda_3) = 1$, $\mathrm{Inf}(B, \Lambda_1, \Lambda_3) = 0$, and $\mathrm{Inf}(B, \Lambda_2, \Lambda_3) = 0$ (as always we write 1 for *True* and 0 for *false*), i.e. $\Lambda_1 \nrightarrow \Lambda_3$, and $\Lambda_2 \nrightarrow \Lambda_3$. Then at least one of the predicates $\widetilde{\mathrm{Inf}}(B, \Lambda_1, \Lambda_3)$ and $\widetilde{\mathrm{Inf}}(B, \Lambda_2, \Lambda_3)$ is true, say the first, and one of the intersections of the boards $T_{\sigma_1}(\lambda_1') \cap T_{\sigma_2}(\lambda_2)$ that define it is not empty. The corresponding intersection $T_{\sigma_1}(\lambda_1) \cap T_{\sigma_2}(\lambda_2)$, which enters the expression defining the influence predicate $\mathrm{Inf}(B, \Lambda_1, \Lambda_3)$, is empty. Therefore $T_{\sigma_1}(\lambda') = N(\lambda_1', \mu)$ for some long range piece μ of color col B, and $T_{\sigma_2}(\lambda_2)$ is the board $E(\Theta)$ for some move $\Theta \in \Lambda_3$ or the board $I(\Theta)$ for a move $\Theta \in \Lambda_3$ of color opposite to col B. In either case the board marks only a single square $p \in N(\lambda_1', \mu)$, $p \notin N(\lambda_1, \mu)$.

Since $\lambda_1' \in \Lambda_1 + \Lambda_2$, the branch $\Lambda_1 + \Lambda_2$ is a strict composition of the branches Λ', λ'_1, and Λ'', i. e.

$$\Lambda = \Lambda' * \lambda_1' * \Lambda''.$$

The branches Λ' and Λ'' consist of moves in the branches Λ_1 and Λ_2; the Λ_1 moves precede the Λ_2 moves in the corresponding branches Λ' and Λ'', while the moves belonging to λ'_1 belong to the branch Λ_1 and occur in it between moves belonging to Λ' and Λ''. Thus

$$\Lambda' = \Lambda_1' + \Lambda_2',$$
$$\Lambda' * \lambda_1' = \Lambda_1' * \lambda_1 + \Lambda'_2,$$

where Λ_1' and $\Lambda'_1 * \lambda_1$ are initial portions of the branch Λ_1, and Λ'_2 is an initial portion of Λ_2.

We consider first the case in which the first move Ψ' of the two-move branches $\lambda_1(A; \Psi', \Psi'') \in \Lambda_1$ and $\lambda'(A'; \Psi', \Psi'') \in \Lambda_1 + \Lambda_2$ is a move by the piece μ, i. e. $\varphi(\Psi') = \mu$. Then

$$p \in N(\lambda', \mu) = P(\text{fin}(B, \Lambda' * \lambda_1'), \mu)$$
$$p \notin N(\lambda_1, \mu) = P(\text{fin}(B, \Lambda_1' * \lambda_1), \mu)$$

and the piece μ occupies one and the same square (Ψ') in the positions $\text{fin}(B, \Lambda' * \lambda'_1)$ and $\text{fin}(B, \Lambda_1' * \lambda_1)$. This means that in the second position the trajectory from the square (Ψ') to the square p is blocked by pieces. The blocking piece nearest to $I(\Psi')$ is on some square $p' \in P(\text{fin}(B, \Lambda_1' * \lambda_1), \mu) = N(\lambda_1, \mu)$, which is freed by some move $\Theta_n \in \Lambda_1'$. Therefore $E(\Theta_n) = p'$ and the square p' is marked in the intersection $N(\lambda_1, \mu) \cap E(\Theta_n)$. This intersection is therefore not empty, so that $\Lambda_1 \sim \Lambda_2$.

Now let $\varphi(\Psi') \neq \mu$. Then the piece μ occupies one and the same square q in the positions $\text{fin}(B, \Lambda')$ and $\text{fin}(B, \Lambda' * \lambda_1')$, while

$$p \in N(\lambda_1', \mu) = P(\text{fin}(B, \Lambda' * \lambda'_1), \mu) \setminus P(\text{fin}(B, \Lambda'), \mu).$$

That is to say

$$p \in P(\text{fin}(B, \Lambda' * \lambda'_1), \mu),$$

i.e. the position $\text{fin}(B, \Lambda')$ contains at least one piece on the trajectory between the squares q and p. The nearest of these (possibly there is only one) left its square by the move Ψ' or Ψ''. The piece μ is on the square q either in one of the positions $\text{fin}(B, \Lambda'_1)$ or $\text{fin}(B, \Lambda'_2)$ or in both. Let us examine the two cases.

If the piece μ is on the square q in only one of these positions, it arrived there by one of the moves $\Theta_{2\varepsilon}$ in the corresponding branch Λ'_1 or Λ'_2 (the proof of this assertion is based, as above, on a count of the number of the arrivals and departures of μ at q in the branch $\Lambda'_1 + \Lambda'_2$). Since the piece does not move in the branch λ_1 ($\varphi(\Psi') \neq \mu$), it occupies the same square in the positions $\text{fin}(B, \Lambda'_1 * \lambda_1)$ and $\text{fin}(B, \Lambda'_1)$ (it cannot have been captured in the move Ψ'', else it would not appear in $\text{fin}(B, \Lambda' * \lambda_1)$). Therefore, in that one of the positions $\text{fin}(B, \Lambda'_1 * \lambda_1)$, $\text{fin}(B, \Lambda'_2)$ in which it occupies the square q, all the squares attacked by it belong to one of the new-possibilities boards $N(\lambda, \mu)$, where the two-move branch λ belongs to $\Lambda'_1 * \lambda_1$ or Λ'_2, respectively. Since $\Lambda_1 \nsim \Lambda_3$ and $\Lambda_2 \nsim \Lambda_3$, the square p appears on none of these boards.

Therefore the square p is not attacked by the piece μ in that one of the positions $\operatorname{fin}(B, \Lambda'_1 * \lambda_1)$, $\operatorname{fin}(B, \Lambda'_2)$ where μ is on the square q. The trajectory between the squares p and q is blocked in that position. Let p' be the square nearest q occupied by a blocking piece. It is attacked by the piece μ and, as we showed earlier, belongs to one of the boards $N(\lambda, \mu)$, where the branch λ belongs to one of the branches $\Lambda'_1 * \lambda_1$, Λ'_2. The square p' is freed by a move in the other branch (Λ'_2 or $\Lambda'_1 * \lambda_1$). Therefore $N(\lambda, \mu) \cap E(\Theta_n) \neq \varnothing$, and so $\Lambda_1 \frown \Lambda_2$.

Finally, let μ be on the square q in both the positions $\operatorname{fin}(B, \Lambda'_1)$ and $\operatorname{fin}(B, \Lambda'_2)$, and therefore in the position $\operatorname{fin}(B, \Lambda'_1 * \lambda_1)$. At the move Ψ' or Ψ'' in the branch $\Lambda'_1 * \lambda_1$ some piece leaves the trajectory between the squares q and p. But $p' \in N(\lambda_1, \mu)$, so that the trajectory remains blocked. The blocking pieces set it free by moves $\Theta_n \in \Lambda'_2$, since it is free in the position $\operatorname{fin}(B, \Lambda' * \lambda'_1)$. That is to say, our trajectory is liberated by the 'combined forces' of moves in the branches $\Lambda'_1 * \lambda_1$ and Λ'_2. During the construction of the influence predicate it was shown that in this case $N(s, \Lambda'_1 * \lambda_1) \cap (E(s, \Lambda'_2) \cup E(c, \Lambda'_2)) \neq \varnothing$ or $N(s, \Lambda'_2) \cap (E(s, \Lambda_1 * \lambda_1) \cup E(c, \Lambda'_1 * \lambda_1)) \neq \varnothing$. Thus $\Lambda'_1 * \lambda_1 \frown \Lambda_2$, whence it follows that in our current case we have $\Lambda_1 \frown \Lambda_2$.

An immediate consequence of the construction of the influence predicate $I(B, L_1, L_2)$ is that the axiom on Situation 1 and the first axiom on Situation 2 are satisfied. We showed in the construction that the conditions of the second axiom on Situation 2 never hold (hence the axiom is formally satisfied). Therefore in the proofs of the theorems on the transfer of scores we may omit the cases in which it is applied, and so dispense with the influence of branches on a move and the corresponding axioms. We may however introduce another concept of equivalent moves, which helps in reducing the number of different virtual moves. The different moves of a long range piece that all take place in a single rank, file, or diagonal and all lead to the same square may be taken as equivalent. Then the satisfaction of the second axiom on Situation 2 becomes possible, and we may then so define the predicate for the influence of branches on a move that the second axiom on Situation 2 and the axioms on the influence of branches on a move are satisfied.

The axiom on Situations 3 and 4 is needed only when (a) a branch Λ_2 whose end is a terminal position is of odd length and the position $\operatorname{fin}(B, \Lambda_2)$ is lost for col B, or (b) the end of the branch $\Lambda_1 + \Lambda_2$, of even length, is terminal and not lost. We may suppose that in our game such positions do not exist. In fact, a complete loss for either side occurs in a position where the losing side is to move after the opponent has captured the King on the preceding move (if a loss occurs in obtaining a mate, it can occur only in positions where one has the move). Since our game does not admit of a stalemate, a terminal position with a drawn score is a conditional concept. We may therefore suppose that our opponent will agree to a draw only in positions where the move belongs to the side that considers a draw a loss for itself. Then the terminal positions $\mathrm{fin}(B, L_1)$ and $\mathrm{fin}(B, L_1 + L_2)$ are either both lost or both not lost. Accordingly, we need not apply the axiom on Situation 5.

Thus in our game both the theorems proved earlier, on the transfer of scores, are valid. If, however, we are to consider all the rules of chess, we must include in our construction of an influence predicate boards connected with checks. Then some of the influence axioms will fail to be satisfied in some cases and we must supplement the theorems on the transfer of scores by an investigation of such special cases. We introduce without proof the construction of the influence predicate when checks are taken into account. We define the following supplementary boards:

(1) $C(\Psi)$ is a trajectory in which the move Ψ gives check (if it gives check at all), including the square occupied by the checking piece and the square occupied by the checked King;
(2) $C'_{\mathrm{w,b}}(\Psi)$ is a trajectory in which a pin arises (if it arises at all) including the squares containing the pinning and pinned pieces and the King (White or Black) of the color of the pinning piece;
(3) $\overline{W}(\Psi) := W(\Psi) \cup C(\Psi) \cup C'_{\mathrm{col}\,\Psi}(\Psi)$;
(4) $N_c(\lambda)$ represents the new possibilities of giving check to the King of color opposed to $\mathrm{col}\,\lambda$, i.e. the squares from which King can be attached by an opposing piece not yet removed from the board after moves in the branch λ and not attacked before those moves

$$S(\lambda(A;\Psi',\Psi''),\mu) := P(A,\mu) \backslash (P(\mathrm{fin}(A,\lambda),\mu)$$

(5) $R(\Psi)$ are the squares occupied by opposing pieces around the opponent's King, if the move Ψ is the check;
(6) $Q(\Psi)$ are the squares around the King checked by the move Ψ that are attacked by pieces of color $\mathrm{col}\,\Psi$.

The new boards corresponding to moves and two-move branches belonging to the branch under investigation or to the subtree $\mathfrak{L}$ will also be integrated along with the pieces of the given color in positions belonging to $\mathfrak{L}$. We shall need the following boards:

$$\overline{W}(c,\mathfrak{L}) := \bigcup_{\Psi \in \mathfrak{L},\, \mathrm{col}\,\Psi \neq \mathrm{col}\,B} \overline{W}(\Psi);$$

$$C(c,\mathfrak{L}) := \bigcup_{\Psi \in \mathfrak{L},\, \mathrm{col}\,\Psi \neq \mathrm{col}\,B} C(\Psi);$$

$$N_c(s,\mathfrak{L}) := \bigcup_{\lambda \in \mathfrak{L},\, \mathrm{col}\,\Psi = \mathrm{col}\,B} N_c(\lambda);$$

$$S(c,\mathfrak{L}) := \bigcup_{\lambda \in \mathfrak{L},\, \mu \in \phi \cup \Pi,\, \mathrm{col}\,\lambda = \mathrm{col}\,\mu \neq \mathrm{col}\,B} S(\lambda,\mu);$$

$$R(s,\mathfrak{L}) := \bigcup_{\Psi \in \mathfrak{L},\, \mathrm{col}\,\Psi \neq \mathrm{col}\,B} R(\Psi);$$

$$Q(s,\mathfrak{L}) := \bigcup_{\Psi \in \mathfrak{L},\, \mathrm{col}\,\Psi \neq \mathrm{col}\,B} Q(\Psi);$$

$$\overline{N}(s,\mathfrak{L}) := N(s,\mathfrak{L}) \cup n_p(s,\mathfrak{L}) \cup n_b(s,\mathfrak{L}).$$

The influence predicate is given by the formula

$$
\begin{aligned}
\mathrm{Inf}(B, L_1, L_2) :=\\
&= (((E(s, L_1) \cup E(c, L_1) \cup I(c, L_1)) \cap (E(c, L_2) \cup I(s, L_2)\\
&\quad \cup I(c, L_2) \cup \overline{W}(c, L_2) \cup \overline{N}(s, L_2))) \cup (I(s, L_1) \cap \overline{W}(c, L_2))\\
&\quad \cup ((E(s, L_1) \cup E(c, L_1)) \cap (N_c(s, L_2) \cup R(s, L_2))) \cup (N_c(s, L_1)\\
&\quad \cap (E(s, L_2) \cup E(c, L_2) \cup \overline{N}(s, L_2)))\\
&\quad \cup ((I(s, L_1) \cup I(c, L_1) \cup R(s, L_1)) \cap (E(s, L_2) \cup E(c, L_2)))\\
&\quad\quad \cup (\overline{W}(c, L_1) \cap (E(s, L_2) \cup E(c, L_2) \cup I(s, L_2) \cup I(c, L_2)))\\
&\quad\quad \cup (\overline{N}(s, L_1) \cap (E(s, L_2) \cup E(c, L_2)\\
&\quad\quad \cup I(c, L_2) \cup C(c, L_2) \cup N_c(c, L_2))) \cup (C(c, L_1) \cap \overline{N}(s, L_2))\\
&\quad\quad \cup (S(c, L_1) \cap Q(s, L_2)) \cup (Q(s, L_2)\\
&\quad\quad \cap S(c, L_1)) \cup G(B, L_1, L_2).
\end{aligned}
$$

Let us now pass to the construction of the influence predicate for the game of noughts-and-crosses. We note first that situations of type 2 cannot occur in this game. For, let Λ_1 and Λ_2 be branches of even length, and let $\Lambda = \Lambda_1 + \Lambda_2$. In the position $\mathrm{fin}(B, \Lambda_1 + \Lambda_2)$ the side corresponding to col B can place its mark (say a nought) on any free square and only on such a square. But the free squares in this position are also free in the positions $\mathrm{fin}(B, \Lambda_1)$ and $\mathrm{fin}(B, \Lambda_2)$. At the end of the composite branch $\Lambda_1 + \Lambda_2$ only those moves are admissible that were admissible at the ends of the constituent branches Λ_1 and Λ_2. Thus there remains for consideration Situation 1.

Let L_1 be a branch with odd length, L_2 a branch with even length. Suppose that in the position $\mathrm{fin}(B, L_1)$ a cross may be marked in the square p, but cannot be marked there in the position $\mathrm{fin}(B, L_1 + L_2)$. This may happen for one of the following reasons:

(1) The square p contains a nought, which must have been placed there in a move $\Theta \in L_2$.

(2) The square p, in which the cross is to be marked, already contains a cross, which must have been put there in some move $\Theta_{2\varepsilon} = \Psi_p \in L_2$. In this case we do not suppose that $L_1 \sim L_2$. In fact, in constructing a pseudocritical branch we make the move Ω_q to a free square q from the position $\mathrm{fin}(B, L_1 + L_2)$ with move belonging to the opponent of col B. By the inductive assertion this position is not lost; the beginning of the induction is the assertion that the position $C = \mathrm{fin}(B, L_2)$ is not lost. The position $\mathrm{fin}(B,(L_1 + L_2) * \Omega_q)$ is not lost since no move from $\mathrm{fin}(B, \Lambda_1 + \Lambda_2)$ wins. This position coincides with the position $\mathrm{fin}(B, L_1 * \Xi_q + L_2')$, where L_2' is the branch in which the move $\Theta_{2\varepsilon} = \Psi_p$ is replaced by the move $\Theta_{2\varepsilon'} = \Omega_q$ (see Figure 33). Such a replacement is permissible since the square q, which is free in the position $\mathrm{fin}(B, L_1 + L_2)$ is free in all the preceding positions in the branch $L_1 + L_2$.

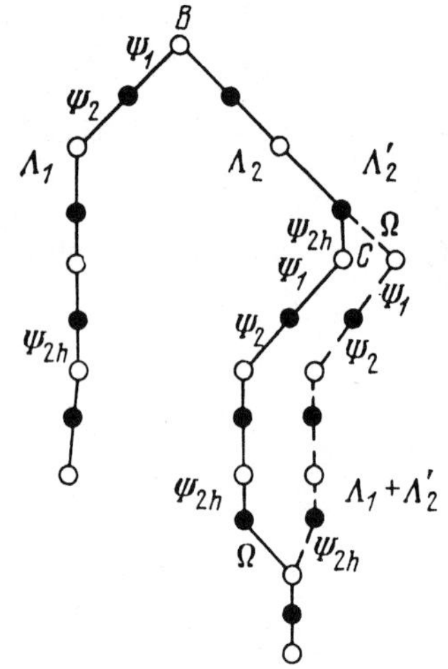

Figure 33

(3) No move can be made from the position fin$(B, L_1 + L_2)$ since it is a terminal position, that is, the side of color col B completed a quintuple in the preceding move. The quintuple consists of three sets of noughts: those that were in place in the position B and those that were placed by moves in the branches L_1 and L_2, these latter sets being non-empty. The branch L_2 can be distinguished from the initial branch $L_2^{(0)}$ with which we began the construction of the pseudocritical branch, but we do not change the moves of color col B. That is, the quintuple consists of noughts from the branches L_1 and $L_2^{(0)}$ and from the position B.

We arrive at the following definition of the influence predicate Inf$(B, L_1, L_2))$: Let $T^0(B)$ be a board on which the noughts are in the same places as in B; let $T^0(L_1)$ and $T^0(L_2)$ be boards for the noughts placed by moves in the branches L_1 and L_2: and let $T^\times(L_1)$ be the board for crosses placed by moves in the branch L_1. Then

$$\mathrm{Inf}(B, L_1, L_2) := \left[T^\times(L_1) \cap T^0(L_2)\right]$$
$$\cup \,\mathrm{Pent}\left(T^0(B) \cup T^0(L_1) \cup T^0(L_2)\right);$$

where Pent(T) is a predicate with the value 1 if the board T contains a quintuple of squares (in a horizontal, vertical, or diagonal line), and with the value 0 otherwise. The influence predicate so defined yields a natural definition of the influence of a branch on a branch or subtree.

We have not symmetrized this influence predicate, and have not added the requirement that the second axiom on the composition of branches should be satisfied. (The first axiom is not needed since we have avoided the influence of a branch on a move. For an arbitrary branch L of even length, only those moves are admissible from the position fin(B, L) that were admissible from the position B.) We can therefore manage without (a) the theorem on the transfer of scores of groups of moves and (b) the resolution of a pseudocritical path into its components. The axioms in question are not used in the proof of the theorem on the transfer of a position score (nor in the similar proof of the theorem on the transfer of the score of a move).

Influence-Based Algorithms

All of the algorithms described below have been implemented in the chess program KAISSA and in this section we shall therefore discuss only the game of chess.

The first algorithm prunes subtrees in the forced game. We may think of the forced game as an auxiliary search from a terminal position in some model game with the aim of producing a dynamic score for the position since a static score may be unreliable.

For instance, a position F in which the Queen of color opposite col F is under attack, cannot be evaluated without taking into account the consequences of the capture. Therefore, to obtain a reliable score we make an auxiliary search in which we recognize only captures, checks, and replies to a check. In the preceding chapter we called such a model a forced game.

Forced variations are searched in all terminal positions and almost always in nearby positions they are identical. The theorem on the transfer of position scores holds for a forced game and for the constructively defined influence relation developed in the preceding section (in essence, this was proved in Section 2 of this chapter, where situations connected with checks were not considered). Suppose that we have produced a forced game from the position B and it turns out that it yields a loss. Then from the position $C = \text{fin}(B, L)$ with col B having the move, the forced game also loses if the following conditions are satisfied:

in the forced game the branch L does not influence the minimally pruned subtree of color opposite to col B from the position B;
no new forcing moves can be made from the position $C = \text{fin}(B, L)$.

Thus in many positions we need not run a forced variation. The effectiveness of the algorithm using the result of the theorem on the transfer of position scores depends in large measure on a successful choice of the position B. In a non-terminal position B a trial forced variation is run with the sole aim of applying the results of the search to the terminal positions in the B-subtree.

In choosing B we encounter conflicting requirements. On the one hand, it should be near the root of the tree so that its subtree will contain many terminal positions. On the other hand, we need a small value for the probability that the branches stemming from B and leading to the terminal position C will influence the trial forced game. Therefore the position B should be near the final positions, i. e. should be rather far from the root of the tree. Moreover, if it turns out that the trial forced game from B does not lose, the time spent in running it will have been wasted.

In the implemented algorithm for a search to a fixed depth n the trial forced games are run from positions of depth $n-2$ for which the static score is unsatisfactory. This criterion increases the probability of a loss for the forced variations from terminal positions of depth n.

The depth $n-2$ is maximal for the test models $\mathfrak{S}_T(B)$. The branches L from B to C, where their results are applied, have length $d(L)=2$. However, an increase in the number of trial model games to be investigated is often compensated by the possibility of pruning branches in the base model. If the position B is won for its own color $\mathrm{col}(B)$ in the game $\mathfrak{S}_T(B)$, it is won in the base model. Then we may step back from B without investigating the quiet moves in the base-model B-subtree.

In fact, if quiet moves are admissible in the base model at a position D which arises after a winning move $\Psi=(B, D)$ in the forced game, then a blank move $m_\varnothing$ from D is admissible in the game $\mathfrak{S}_T(B)$. Since it loses, the material balance at D favors col B. After a quiet move $\Theta=(D, F)$ from the position D the position F has the same material balance and is terminal in the base model. At F we consider only moves in the forced game, while the blank move $m_\varnothing$ is admissible (else $\Theta=(D, F)$ would not be a quiet move) and it wins. Moves from D in the forced game were investigated in the trial model game $\Sigma_T(B)$; they also lead to positions in the base model from which only moves in the forced game are considered. These have been investigated in the trial model $\Sigma_T(B)$, and some of them win. Therefore B is a won position for col B in the base model.

The fundamental defect in this algorithm is its instability: if a new forced move appears at C, the forced variation must be repeated. The algorithm does not recognize situations in which the new moves bear no relation to the trial forced game.

On close examination the forced games turn out not to be identical, but instead similar. More exactly, there exists a set of fragments—subtrees of the base tree such that in almost every position of the forced game we find a union of a subset of these fragments. However this subset may be unique for each position.

It is natural to examine influence with respect to fragments. There are two different algorithms using such methods for shortening the search. The second is based on theory; we shall turn to it later.

We begin with an algorithm equipped with a table of poor moves. At every point in the search the table of poor moves contains fragments assigned to the various positions in the current branch. In fact the table contains the first move in the fragment, the opponent's best reply, the amount of material loss sustained in the fragment, and the boards belonging to the fragment.

We define the material loss for a poor move as the difference between the material score at the position A as seen by the side with color col A and the material score at a position reached from A as a result of that move. Thus if the move Ψ loses a minor piece, its material loss is equal to the weight h_{mp}. If the move Ψ loses no more than a Pawn, the material loss is equal to h_P. Nevertheless the move Ψ may be poor if there is reason to try for greater gain. All the moves in the fragment, except possibly the first, are forced. The fact that the fragment is assigned to a position in the current branch implies that in this position the fragment would also run its course. It is not at all necessary that the fragment should in fact have been generated in this position. Moreover, as we shall see, it is not necessary that the fragment should have been generated in any position whatever. It may have been put together from smaller fragments that were generated in various positions. We know only that it is assembled from the subtree of the position to which it is assigned.

Fragments with blank first moves occupy a special place. They characterize winning forced variations; the need for them arises in attempts to put together a strong fragment from weak ones. They have, however, a meaning of their own. Assigned to a node of opposing color they point to the possibility of a forced win for the side having the move: it is for this reason that they are called threats.

Fragments that define threats or, what amounts to the same thing, fragments with a blank first move, will be called *k^0-threats* and the remaining fragments will be called *l^0-poor moves*. The color of an l^0-poor move is the color of the side that opens the fragment. The color of a k^0-threat is defined quite naturally as the color of the side making the blank move, e. g. a 'White k^0-threat' is a threat to White.

The formula developed in the preceding section, for the value of the influence predicate $I(B, L_1, L_2)$, can be used to determine the mutual influence of branches and subtrees stemming from different nodes. However, our observations on the usability of such influence predicates have only a heuristic character.

The position B to which a fragment is assigned is regarded as the root of the corresponding subtree of the game. In practice, however, the fragment may be formed during the traversal of a subtree having its root at another node. It is transferred to the position B along some branch on which it has no influence. Figure 34 depicts a portion of a game tree. Certain subtrees in this portion are distinguished (by the fact that they lead to a change in the material balance). They represent k^0-threats and l^0-poor moves assigned to the position D. After a traversal of the D-subtree of the game $\mathfrak{A}$ they are all assigned to the position D. After a backward step from D some of them in a transformed version will be assigned to the preceding position B.

The traversal algorithm consists of forward and backward steps; we first consider the elaboration of fragments during a backward step.

Some of the fragments assigned to D are assembled to form fragments that may be transferred to B. Other fragments are merely lifted from the table. Suppose the move $\Psi = (B, D)$ leads from B to D. We map it into

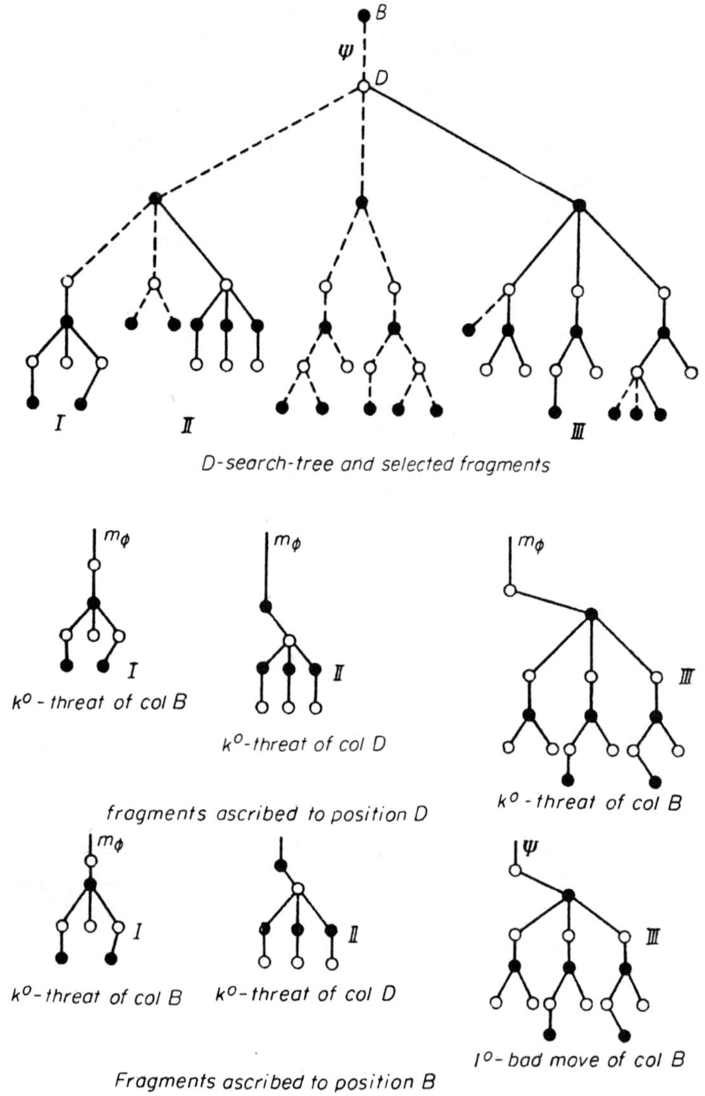

Figure 34

two (formally) two-move branches $\lambda_1(B; \Psi, m_\varnothing)$ and $\lambda_2(B; m_\varnothing, \Psi)$. The origin of the first is the position B; the origin of the second has the same configuration of pieces as B, but the move belongs to the opposite side.

We test each k^0-threat for its influence on the branches λ_1 and λ_2 of its own color: if $\operatorname{col} k^0 = \operatorname{col} \Psi$, the influence on λ_1, else on λ_2. If there is no influence we test again to see whether the threat was first obtained from the minimax portion of the search tree from D (the union of the minimal

White- and Black-pruned subtrees). We get a negative answer in two cases: $\mathrm{col}\, k^0 = \mathrm{col}\, \Psi$ and Ψ is a poor move, or $\mathrm{col}\, k^0 \neq \mathrm{col}\, \Psi$ and Ψ is closing. In both cases it must be remembered that the threat in question did not arise in the minimax portion of the game tree. We call it a marked threat; once marked, a threat remains marked until it is excluded from the table.

If a k^0-threat of color $\mathrm{col}\, k^0 = \mathrm{col}\, \Psi$ does not influence the branch λ_1, or if a k^0-threat of color $\mathrm{col}\, k^0 \neq \mathrm{col}\, \Psi$ does not influence the branch λ_2, it is assigned to the position B. If it does influence the corresponding branch, however, we proceed differently: k^0-threats of color $\mathrm{col}\, k^0 \neq \mathrm{col}\, \Psi$ and marked threats k^0 of color $\mathrm{col}\, k^0 = \mathrm{col}\, \Psi$ are excluded from the table, and instead of unmarked k^0-threats of color $\mathrm{col}\, \Psi$ we form l^0-poor moves and assign them to B. The first move $\Theta_0(l^0)$ of the new l^0-poor move, the opponent's best reply $\Theta_1(l^0)$, the material loss $\mathrm{Mat}(l^0)$ and the ensemble of boards $\{T_\gamma(l^0)\}$ are defined in the following way:

$$\Theta_0(l^0) := \Psi;$$
$$\Theta_1(l^0) := \Theta_1(k^0);$$
$$\mathrm{Mat}(l^0) := \mathrm{Mat}(k^0) - h(\Psi);$$
$$\{T_\gamma(l^0)\} := \{T_\gamma(k^0)\} \cup \{T_\gamma(\lambda_1)\}.$$

Here $h(\Psi)$ is the weight of the piece captured in the move Ψ.

Let us see how we process l^0-poor moves. At their outset we form an ensemble of empty boards of the so-called summator, in which we later form the logical union of the boards of some of the contemplated l^0-poor moves assigned to the position D. For each of the l^0-poor moves we test to see whether it influences the branch λ_1 or λ_2 of its own color; if it has no influence, it is assigned to B and is marked in the same way as a k^0-threat.

If an l^0-poor move of color $\mathrm{col}\, B$ influences the branch λ_1, it is excluded from the table. We also exclude l^0-poor moves of color $\mathrm{col}\, l^0 \neq \mathrm{col}\, \Psi$ that influence the branch λ_2 when Ψ is a poor move. The boards for the remaining l^0-poor moves that influence the branch λ_2 are combined, i. e. every set of such boards is in turn combined with the set of boards in the summator. In addition, we determine the minimum material loss for all such l^0-poor moves.

We now form the fragments assigned to the position B. If the move Ψ is poor and some k^0-threats of color $\mathrm{col}\, B$ appear on the branch λ_1, we form a new l^0-poor move:

$$\Theta_0(l^0) := \Psi;$$
$$\Theta_1(l^0) := m_\varnothing;$$
$$\mathrm{Mat}(l^0) := -h(\Psi);$$
$$\{T_\gamma(l^0)\} := \{T_\gamma(\lambda_1)\}.$$

We have no further information about the new l^0-poor move.

If the move Ψ is a refutation, we construct the corresponding k^0-threat, of color opposite to $\operatorname{col}\Psi = \operatorname{col} B$:

$$\Theta_0(k^0) := m_{\varnothing};$$
$$\Theta_1(k^0) := \Psi;$$
$$\operatorname{Mat}(k^0) := \min_{\operatorname{col} l^0 \neq \operatorname{col}\Psi,\ l^0 \sim \lambda_2} \operatorname{Mat}(\lambda_0) + h(\Psi);$$
$$\{T_\gamma(k^0)\} := \{T_\Sigma\}.$$

Here $\{T_\gamma(k^0)\}$ is the set of boards for the new k^0-threat, and $\{T_\Sigma\}$ is the set of boards in the summator. The l^0-poor moves that have influence are excluded from the table.

If Ψ is an improving move the rules that we have stated will correctly construct the k^0-threats and the l^0-poor moves. There may be more than one of the latter, since there may be no k^0-threat that influences the branch λ_1.

After the completion of a forward step the table so constructed is used to shorten the search from the position D whence the step was made. It is used, however, only when nothing but moves in the forced game are considered from the position D. For every k^0-threat of color $\operatorname{col} k^0 = \operatorname{col} D$ we ask whether:

it influences the branch leading to D from the position B to which it is assigned;
it satisfies the inequality

$$f_{\mathrm{m}}(D) \underset{\operatorname{col} D}{\preccurlyeq} \operatorname{Mat}(k^0) + m_{\mathrm{m}},$$

where $f_{\mathrm{m}}(D)$ is the value of the material evaluation function for the position D and m_{m} is the material portion of the bound for the corresponding color ($m_{\mathrm{m}} = \underline{\lim}_{\mathrm{m}}$ if White has the move at D, else $m_{\mathrm{m}} = \overline{\lim}_{\mathrm{m}}$).

If there is a k^0-threat of color $\operatorname{col} B$ with sufficient material gain and no influence on the corresponding branch, and if at the same time there is no k^0-threat of the opposing color, the search is not prolonged beyond D; instead a backward step is made at once, on the grounds that the threat is automatically realized. Thus there is a winning move at the given position.

The table is also used to shorten the search by pruning various moves. After we have generated a move Ψ at D and the controlling algorithm proposes to make it, we test to see whether the table contains l^0-poor moves yielding information about the move Ψ. For every fragment with first move $\Theta_0(l^0) = \Psi$ we ask whether the material loss $\operatorname{Mat}(l^0)$ is sufficient to cause pruning, i. e. whether

$$f_{\mathrm{m}}(D) \underset{\operatorname{col} D}{\preccurlyeq} \operatorname{Mat}(l^0) + m_{\mathrm{m}}.$$

If the material loss is great enough we ask whether the l^0-poor move influences the branch $L(B,\ldots, D)$ leading from the position B to which the move is assigned, to the position D under scrutiny. If we find at least one l^0-poor move with first move Ψ and with sufficient material loss, not influencing the corresponding branch, the investigation of the move Ψ is postponed. At the same time, a note is made of the l^0-poor moves that caused the postponement.

If the investigation of Ψ is not postponed, the fragment with first move Ψ is entered into the table and assigned to the position D. We do not construct any boards, but we note that the fragment is in service. Given a move Ψ at a position in the D-subtree, we first ask if such a fragment exists. If it does we do not test the influence of fragments with first move Ψ assigned to higher positions. Thus we waste no time on tests whose results we already know.

After completing the search in a given position for urgent moves, we test each postponed move to see whether the fragments giving rise to the postponement influence the minimax subtree of the given position. Moves for which all fragments influence that subtree are investigated; the remainder are again postponed. At the end of the search of the admitted moves we again test the postponed moves to see whether their fragments influence the new minimax subtree, and repeat the process as necessary.

This procedure comes to an end in one of two cases. Either we have investigated all the moves, or at some point we have found for each admitted move a fragment that caused its postponement and that does not influence the minimax subtree at the given position. In the latter case all uninvestigated moves are pruned.

Above, we described in principle an outline of an algorithm that has been implemented in a chess-playing program. Several important details were omitted, and for ease of understanding others were presented in a form somewhat different from that which was implemented in the program. For this reason the account that was given above must not be taken as a technical description of the algorithm, the more so because the algorithm is continually being revised and optimized. In particular, in the first version there was no notion of the k^0-threat, which was introduced as an aid in the construction of l^0-poor moves and in the course of time became one of the basic concepts in the algorithm.

The table of l^0-poor moves not only aids in shortening the search; it is also used to improve the order in which moves are investigated and to detect some meaningful circumstances related to the position under scrutiny.

The basic shortcoming in the algorithm considered above is the frequent occurrence of situations in which information is lacking.

A characteristic case occurs when a k^0-threat is not connected with the opponent's preceding move but is connected with the move of one's own that preceded the latter. Then the corresponding k^0-threat cannot be constructed by the method we have just described, since the execution of such a threat requires two successive moves by a single side and is complicated by the opponent's intervening move. On the other hand, an old threat cannot be carried to the top level.

As an example of such a situation we choose a Knight fork against Queen and Rook. Any reply results in the loss of Queen or Rook. These captures are k^0-threats and the opponent's move preceding the capture is not connected with it. What information can be carried back to the level of the position at which the forking move was made? Clearly no k^0-threat existed there, since the forking move is not forced and is not connected with the subsequent move by the opponent. On the other hand, in the pure chess sense, this move is a threat and we have all the information we need about it. Such a threat is simply not comprehended in the scheme of our algorithm.

Let us now bring in the notion of the k^1-threat. A move Ψ is a k^1-*threat* when it gives rise to a set of k^0-threats. A k^1-threat is defined constructively as a move immediately connected to at least one k^0-threat. Then any attack on a piece is a k^1-threat.

Nevertheless, a k^1-threat may be a losing move if it allows the opponent to make a k^0-threat of his own, connected with it. If so, the k^1-threat will be included in the table of l^0-poor moves, on general grounds.

The structure of the information concerning a k^1-threat is as follows: the header contains, in addition to all items conserved for a k^0-threat, the

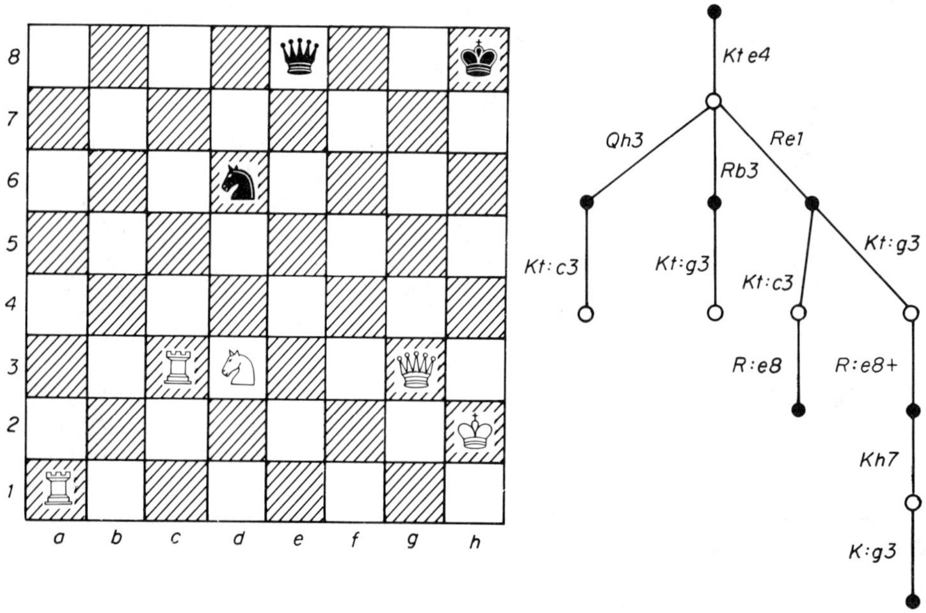

Figure 35

headers for two lists—the list of its own admissible k^0-threats, transferred from the base table of k^0-threats to a special table, and the list of refutation moves by the opponent that are his best replies to the k^1-threat.

The refutations need not be k^0-threats, since they are not forced. They must however influence the k^1-threat. The information on every refutation contains a list of l^0-poor moves of the same color as the k^1-threat and a list of k^0-threats of the opposite color.

Let us look at the information connected with a move that creates a fork. We shall suppose the Knight pinned by an opposing Rook (see Figure 35).

The move *Ne*4 is a k^1-threat because the search for k^0-threats that follows it brings up the moves *N*:*c*3 and *N*:*g*3. The information about these k^0-threats is contained in a special field, inaccessible to the basic algorithm, and retained in a list whose header is contained in the information element describing the move *Ne*4 as a k^1-threat.

The refutation list in this case consists of the single move *Re*1. The list of l^0-poor moves assigned to this refutation consists of the moves *N*:*c*3 and *N*:*g*3 (this list is also contained in the reference table). In this case the list of k^1-threats is empty.

Information about k^1-threats is used in the following way during the search: If a move Ψ contained in the table of k^1-threats is made, all the k^0-threats assigned to it are moved from the reference table to the k^0-threat table, where they will be treated from then on as a common basis. However, after the backward step corresponding to the move Ψ these k^0-threats will be removed from the table without examination.

If in the search a k^1-threat is immediately followed by one of the refutation moves assigned to it, the corresponding l^0-poor moves and k^0-threats belonging to the opponent are transferred to the general table of l^0-poor moves and k^0-threats.

Thus the program does not compel a new search and new construction of these fragments. Not only do we save the time that would have been spent in reconstructing some already constructed (but lost) elements of the table, but we also increase our ability to shorten the search. Thus, in the example of the Knight fork, if the algorithm did not use the information about k^1-threats it would construct the whole tree shown in Figure 35 after the move *Ne*4 and after a backward step corresponding to the move *Ne*4 would lose the information developed during the search of the tree just constructed.

Of course, during the search of the tree it is necessary to verify the correctness of the prior information on the k^1-threat. This is done on general grounds with the aid of the notion of influence. A separate examination is made for the influence of the k^1-threat itself, each of the refutation moves, and each l^0-poor move and each k^0-threat in the reference reserve. In this way the several elements in the list of l^0-poor moves and k^0-threats may be deleted individually. We can also delete refutations together with all of their lists. A k^1-threat itself is eliminated not only when the current move influences it but also when the list of k^0-threats has been emptied by deletions.

When several individual k^1-threats arise at the same level, their lists are edited and merged.

The structure of k^1-threats falls naturally far short of prescribing all the situations in which a precise treatment with k^0-threats is necessary. In our opinion, however, it does reflect certain concepts connected with the notion of 'tempo' or 'time' in several games. On a careful scrutiny of the algorithms we have obtained, it becomes clear that threats are templates for the searchless estimates of trees at level 2, as defined by Botvinnik. In a model of the active game, k^1-threats and l^1-*losing moves* can be used to reduce the tree of the forced game in the test of moves for activity and safety.

One might try to construct algorithms that would work with k^2-threats and others of higher order. It is very easy to define such algorithms formally, but in such games as chess they would require very great amounts of memory for the corresponding reserve tables, while for simpler games it appears that the k^1-threats suffice.

A second algorithm using fragments to shorten the search is based on the notion of a test-correlated variation (TCV). The TCV is constructed recursively; the first step is to construct it at the base position of the search. Here we list all the forced moves and for each move construct a fragment, whose boards initially consist of the board for that move alone. At any move in the search we allow only those of the original list that influence the branch leading from the base position to the move in question. We also list those moves that were originally excluded because the necessary influence was lacking. After inspection and consideration of a move in the search, its fragment is augmented by the boards of the minimax tree in our search. After considering all moves, we determine which ones need no further examination. These are the moves whose fragments influence moves that were excluded from our initial inspection on the grounds that their influence indicated the possibility of an essential change in the results via combination. Moves that win at some stage of the iteration may still be excluded from further inspection. As a result of the successive steps in the iteration the fragments of losing moves will expand, so that multiple inspections of a single move may be required. It can be shown that the number of such iterations does not exceed the number of forced moves at the position. Such a constraint, however, does not forbid the construction of more moves than would be constructed during a search without pruning. Of course, many positions in the tree of an iterative search, which are intentionally much fewer in number, are re-inspected many times.

Nevertheless, even the first application of the iterative search algorithm, in which the iterative search is merely replaced by a customary search of forced variations, will reduce the running time of the program by 10%.

The most recent development of this algorithm involved the creation of a table of fragments in the TCV which, like the k^0-threats and the l^0-poor moves, was carried from one position to another. The boards of fragments from the table were used instead of the boards from an initial move to form the boards of a fragment at the blank iteration.

If a branch leading to a given position from another position where the corresponding TCV was produced does not influence a fragment of the TCV, the latter is treated as the result of the first iteration rather than the result of the blank iteration, i.e. is re-inspected only if some other move influences its fragment.

This focussing of the algorithm led to a rapid convergence of the iteration, and reduced the number of positions in the tree by a factor of 4 or 5.

CHAPTER 4

Algorithms for Games and Probability Theory

Probabilistic Models for Two-Person Games

In Chapter 2 we considered a probabilistic model of a two-person game in which the probability of a correct decision as to the score of the base position is an increasing function of the depth of the search. The results we obtained had a qualitative flavor, inasmuch as their proofs depended on the hypothesis that the values of the evaluation function $f(A)$ for different positions A are independent. It is not clear that this hypothesis is valid; one of the postulates does not hold: The value of the evaluation function $f(B)$ at the position B, following the move $\Psi = (A, B)$ from the position A, depends on the value of $f(A)$. Nevertheless, the method we used (recursive determination of the probability that the model score is correct, as a function of the value of the evaluation function $f(D)$ for positions D of rank $r(D) = n$) can be applied under a much more general set of postulates.

We suppose that the probability characteristics of a position A in our given game $\mathfrak{A}$ are defined by a set of parameters

$$\chi(A) = \chi(t_1(A), t_2(A), \ldots, t_k(A)).$$

The parameters $t_1, t_2, \ldots, t_h$ $(h \leq k)$ depend on the score of the position A, the scores of the positions where branches leading to B originate, the scores of positions reached in branches originating at B, the values of the evaluation function $f(B)$ at the above listed positions, and the color of the move at A. The parameters $t_{h+1}, \ldots, t_k$ have a character independent of the color of the move. These include, for example, the expected number of moves from A, their precision as related to the divergence between the score at A and the true score, etc. The probability characteristics of positions with White or Black to move are symmetric in the sense that if we replace t_i by $1-t_i$ in the set of parameters to which the color of the move pertains, the probability characteristics do not change:

$$\chi(1-t_1,1-t_2,\dots,1-t_h,t_{h+1},\dots,t_k)=$$
$$=\chi(t_1,t_2,\dots,t_h,t_{h+1},\dots,t_k).$$

Among the parameters on which these characteristics depend we single out the turn to move—col A— and we write col $A=0$ for a White position A, col $A=1$ for a Black position. We also single out the true score $\mathrm{sc}(A)$ and the value of the evaluation function $f(A)$. The other parameters $t_4,\dots,t_h$ that change their values in the symmetry formula just as col A does, form a vector $\boldsymbol{q}$; the $t_{h+1},\dots,t_k$ form the vector $\boldsymbol{r}$. The symmetric vector with coordinates $1-t_4,\dots,1-t_h$ will be denoted by $1-\boldsymbol{q}$. Using these symmetries we exclude the color of the turn to move from the explicit parameters and we assume from now on that all the characteristics we deal with relate to a White position. As usual we consider a game $\mathfrak{A}$ in which White and Black move alternately.

We assume that we have a Shannon model or other model whose definition depends on the value of the depth parameter n. In these models we denote the score of the base position A by msc(A). If A is a White position with $\mathrm{sc}(A)=x$, $f(A)=y$, and the remaining probability characteristics having the values $\boldsymbol{q}$ and $\boldsymbol{r}$, we introduce the notation $\mathbf{P}_-(x,y,z,\boldsymbol{q},\boldsymbol{r},n)$ and $\mathbf{P}_+(x,y,z,\boldsymbol{q},\boldsymbol{r},n)$ for the respective probabilities of the events $\mathrm{msc}_n(A)<z$ and $\mathrm{msc}_n(A)\le z$ which may be written out as

$$\boldsymbol{P}_-(x,y,z,q,r,n):=\boldsymbol{P}(\mathrm{msc}_n(A)<z|\mathrm{col}\,A$$
$$=0,\ \mathrm{sc}(A)=x,f(A)=y,\boldsymbol{q}(A)=\boldsymbol{q},\boldsymbol{r}(A)=\boldsymbol{r});$$
$$\boldsymbol{P}_+(x,y,z,q,r,n):=\boldsymbol{P}(\mathrm{msc}_n(A)\le z|\mathrm{col}\,A$$
$$=0,\mathrm{sc}(A)=x,f(A)=y,\boldsymbol{q}(A)=\boldsymbol{q},\boldsymbol{r}(A)=\boldsymbol{r});$$

The inequalities $\mathrm{msc}(A)<\mathrm{z}(\le z)$ hold if and only if the respective inequalities $\mathrm{msc}(B_i)<z(\le z)$ hold for all positions B_i in the corresponding models of depth $n-1$ that are reached by moves $\Psi_i=(A,B_i)$ from A $(i=1,2,\dots,m)$. If the inequalities $\mathrm{msc}_n(A)<z$ or $\mathrm{msc}_n(A)\le z$ hold, so do many similar relationships. From now on we shall treat these all alike, using the symbols $\leqq$, for 'less than' or 'not exceeding'.

Since the B_i are Black positions their characteristics are determined by setting the corresponding parameters to the value $1-z$ (we suppose that the minimum values of the scores and the evaluation functions are equal to zero and the maximum values to one).

We now postulate that for the given depth n the functions $\boldsymbol{P}_-(x,y,z,\boldsymbol{q},\boldsymbol{r},n)$ and $\boldsymbol{P}_+(x,y,z,\boldsymbol{q},\boldsymbol{r},n)$ satisfy the symmetry conditions

$$\boldsymbol{P}_-(x,y,z,\boldsymbol{q},\boldsymbol{r},n)=\boldsymbol{P}(\mathrm{msc}_n(A)>1-z|\mathrm{col}\,A=1,$$
$$\mathrm{sc}(A)=1-x,f(A)=1-y,\boldsymbol{q}(A)=\mathbf{1-\boldsymbol{q}},\boldsymbol{r}(A)=\boldsymbol{r}),$$
$$\boldsymbol{P}_+(x,y,z,\boldsymbol{q},\boldsymbol{r},n)=\boldsymbol{P}(\mathrm{msc}_n(A)\ge 1-z|\mathrm{col}\,A=1,$$
$$\mathrm{sc}(A)=1-x,f(A)=1-y,\boldsymbol{q}(A)=\mathbf{1-\boldsymbol{q}},\boldsymbol{r}(A)=\boldsymbol{r}).$$

These conditions are satisfied for the Shannon model of depth $n = 0$, since in that case $\text{msc}(A) = f(A)$. Then the probabilities that the model score $\text{msc}_0(A)$ will have a given value are independent of all the parameters except $f(A)$ and we have

$$\boldsymbol{P}\left(\text{msc}_0(A) \underset{<}{\leq} z | f(A) = y\right) = \begin{cases} 1, & \text{if } y \underset{<}{\leq} z, \\ 0, & \text{if } y \underset{\geq}{>} z, \end{cases}$$

$$\boldsymbol{P}\left(\text{msc}_0(A) \underset{>}{\geq} 1 - z | f(A) = 1 - y\right) = \begin{cases} 1, & \text{if } 1 - y \underset{>}{\geq} 1 - z, \\ \quad \text{i.e. } y \underset{\leq}{<} z, & \\ 0, & \text{if } 1 - y \underset{\leq}{<} 1 - z, \\ \quad \text{i.e. } y \underset{\geq}{>} z. & \end{cases}$$

This means that the model scores $\text{msc}_n(B_i)$ for the positions arising after the moves $\Psi_i = (A, B_i)$ from A $(i = 1, 2, \ldots, m)$ are mutually independent random variables depending only on the parameters $\text{sc}(B_i) = \xi_i$, $f(B_i) = \eta_i$, $\boldsymbol{q}(B_i) = \gamma_i$, $\boldsymbol{r}(B_i) = \rho_i$. Then the probabilities of the compound events

$$\boldsymbol{P}\left(\left\{\text{msc}_n(B) \underset{>}{\geq} \zeta_i\right\}_1^m | \text{col}\, A = 0, M(A) = \{\Psi_i = (A, B_i)\right.$$
$$\left.\text{sc}(B_i) = \xi_i, f(B_i) = \mu_i, \boldsymbol{q}(B_i) = \gamma_i, \boldsymbol{r}(B_i) = \rho_i\}_1^m\right)$$

are equal to the product of the component probabilities

$$\prod_{i=1}^{m} P_{\pm}(1 - \xi_i, 1 - \eta_i, 1 - \zeta_i, 1 - \gamma_i, \rho_i, n)$$

(we assume that the B_i are Black positions, and we use the symmetry relationships).

With these assumptions and the easily proved relationships

$$\boldsymbol{P}\left(\text{msc}_n(B) \underset{<}{\leq} \zeta | \text{col}\, B = 1, \text{sc}(B) = \xi, f(B) = \eta, \boldsymbol{q}(B) = \gamma,\right.$$
$$\left.\boldsymbol{r}(B) = \rho, n\right) = 1 - P_{\pm}(1 - \xi, 1 - \eta, 1 - \zeta, 1 - \gamma, \rho, n),$$

we derive the recursive equation

$$P_{\pm}(x, y, z, \boldsymbol{q}, \boldsymbol{r}, n + 1)$$
$$= \pi_{\pm}(x, y, z, \boldsymbol{q}, \boldsymbol{r}) + \sum_{m=1}^{M} p(x, y, z, \boldsymbol{q}, \boldsymbol{r}, m)$$
$$\times \int_{\mathfrak{P}_m} \prod_{i=1}^{m} (1 - p_{\pm}(1 - \xi_i, 1 - \eta_i, 1 - z, 1 - \gamma_i, \rho_i, n))$$
$$\times d\Sigma\left(x, y, \boldsymbol{q}, \mathbf{r}, m; \{\xi_i, \eta_i, \gamma_i, \rho_i\}_1^m\right)$$

Here the $\pi_{\pm}(x, y, z, \mathbf{q}, \mathbf{r})$ are the probabilities that a White position A will turn out to be terminal with score not exceeding z (less than z) if $\mathrm{sc}(A) = x$, $f(A) = y$, $\mathbf{q}(A) = \mathbf{q}$, $\mathbf{r}(A) = \mathbf{r}$:

$$\pi_{\pm}(x, y, z, \boldsymbol{q}, \boldsymbol{r}) = \begin{cases} \pi(x, y, \boldsymbol{q}, \boldsymbol{r}), & \text{if } x \underset{<}{\leq} z, \\ 0, & \text{if } x \underset{\geq}{>} z; \end{cases}$$

$p(x, y, \mathbf{q}, \mathbf{r}, m)$ are the probabilities that m moves $\Psi_i = (A, B_i)$ can be made from a White position A if $\mathrm{sc}(A) = x$, $f(A) = y$, $\mathbf{q}(A) = \mathbf{q}$, $\mathbf{r}(A) = \mathbf{r}$; M is the maximum number of moves that can be made from positions in the given game $\mathfrak{A}$. $\mathfrak{P}_m$ is the $k \times m$-dimensional parameter space, the direct product of m k-dimensional parameter spaces defining the probability characteristics of a position:

$$\mathfrak{P}_m = \bigotimes_{i=1}^{m} (X \otimes Y \otimes \mathbf{Q} \otimes \mathbf{R}),$$

where X is the set of possible values of $\mathrm{sc}(A)$, Y is the set of values of $f(A)$, and $\boldsymbol{Q}$ and $\boldsymbol{R}$ are the sets of values of $\boldsymbol{q}$ and $\boldsymbol{r}$. We write $\Sigma(x, y, \mathbf{q}, \mathbf{r}, m; \Omega)$ for the completely additive set function $\Omega \subset \mathfrak{P}_m$ equal to the conditional probability that the set of parameters $\{\mathrm{sc}(B_i) = \xi_i,\ f(B_i) = \eta_i,\ \mathbf{q}(B_i) = \boldsymbol{\gamma}_i,\ \mathbf{r}(B_i) = \boldsymbol{\rho}_i\}_1^m$ defines a point in the space $\mathfrak{P}_m$ lying in the set Ω when $\mathrm{sc}(A) = x$, $f(A) = y$, $\mathbf{q}(A) = \mathbf{q}$, $\mathbf{r}(A) = \mathbf{r}$, and m is the number of moves $\Psi_i = (A, B_i)$ admissible under the rules of the game $\mathfrak{A}$.

Our recursion equation implies that the symmetry property is hereditary, in the sense that if the probabilities $P_{\pm}(x, y, z, \mathbf{q}, \mathbf{r}, n)$ and the characteristics $\pi(x, y, \mathbf{q}, \mathbf{r})$, $p(x, y, \mathbf{q}, \mathbf{r}, m)$ and $\Sigma(x, y, \mathbf{q}, \mathbf{r}, m; \Omega)$ satisfy the symmetry conditions, then so do the probabilities $P_{\pm}(x, y, z, \mathbf{q}, \mathbf{r}, n+1)$. We shall be considering some more or less simple probability models that provide concrete definitions of the probability characteristics and we discuss the relationship these models bear to real games and the evaluation functions that can be constructed for them. We shall suppose that the game $\mathfrak{A}$ is completely uniform (i.e. the number m of moves at a non-terminal position A is the same for all such A) and that the terminal positions have a very high rank $N \gg n$ (consequently $\pi(x, y, \mathbf{q}, \mathbf{r}) = 0$).

We shall normally assume that the scores in any given position have only two values, 1 and 0. When there are intermediate values, we can reduce the situation to this case. For instance, if we wish to know whether White can obtain the score z, we may replace the score $\mathrm{sc}(A)$ in the game $\mathfrak{A}$ by the threshold function

$$\mathrm{sc}_z(A) := \begin{cases} 1, & \text{if } \mathrm{sc}(A) \geq z; \\ 0, & \text{if } \mathrm{sc}(A) < z. \end{cases}$$

When the evaluation function $f(A)$ can take any value between 0 and 1, and the functions $P_+(x, y, z, \mathbf{q}, \mathbf{r}, n)$ and $P_-(x, y, z, \mathbf{q}, \mathbf{r}, n)$ coincide, we may use the notation $P(x, y, z, \mathbf{q}, \mathbf{r}, n)$:

$$P(x, y, z, \boldsymbol{q}, \boldsymbol{r}, n) = P(\mathrm{msc}_n(A) \leq z | \mathrm{col}\, A = 0, \mathrm{sc}(A) = x,$$
$$f(A) = y, \boldsymbol{q}(A) = \boldsymbol{q}, \boldsymbol{r}(A) = \boldsymbol{r})$$
$$= P(\mathrm{msc}_n(A) < z | \mathrm{col}\, A = 0, \mathrm{sc}(A) = x, f(A)$$
$$\leq y, \boldsymbol{q}(A) = \boldsymbol{q}, \boldsymbol{r}(A) = \boldsymbol{r}).$$

Let us return to the model used in Chapter 2. The evaluation function $f(A)$ and the true score $\mathrm{sc}(A)$ take on the values 0 and 1, and the probability characteristics for the position A depend only on the turn to move, col A, and the true score $\mathrm{sc}(A)$. If our given game $\mathfrak{A}$ is completely uniform, with m moves at each non-terminal position, and if at each non-terminal position won for the side having the move there are s winning moves, our general recursion equation reduces to the one considered in Chapter 2:

$$P_+(1, 0, n+1) = 1 - P_{n+1} = (1 - P_+(0, 0, n))^s (1 - P_+(1, 0, n))^{m-s}$$
$$= Q_n^s P_n^{m-s},$$
$$P_+(0, 0, n+1) = 1 - Q_{n+1} = (1 - P_+(1, 0, n))^m = P_n^m.$$

These equations yield the probability P_n that the model score $\mathrm{msc}_n(A)$ is correct for a position A which is won for its own color, and the probability Q_n of an error in the model score for a position A lost for its own color. We need to know only the probabilities for the values of the evaluation function $f(A)$:

$$P_0 = p = \boldsymbol{P}(f(A) = 1 | \mathrm{col}\, A = 0,\ \mathrm{sc}(A) = 1),$$
$$Q_0 = q = \boldsymbol{P}(f(A) = 1 | \mathrm{col}\, A = 0,\ \mathrm{sc}(A) = 0).$$

According to our fundamental assumptions these probability characteristics satisfy the symmetry conditions

$$p = \boldsymbol{P}(f(A) = 0 | \mathrm{col}\, A = 1,\ \mathrm{sc}(A) = 0),$$
$$q = \boldsymbol{P}(f(A) = 0 | \mathrm{col}\, A = 1,\ \mathrm{sc}(A) = 1).$$

In the space of possible values of P_n and Q_n $(0 \leq P_n \leq 1, 0 \leq Q_n \leq 1)$ there are three stationary points, the solutions of the system of equations

$$1 - P = Q^s P^{m-s},$$
$$1 - Q = P^m$$

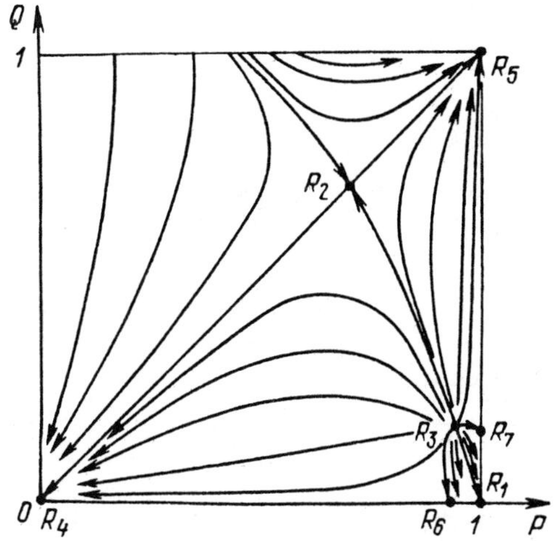

Figure 36

One of these, $R_1(1,0)$ corresponds to the case in which the value of the evaluation function $f(A)$ coincides with the true score $\mathrm{sc}(A)$. Another, $R_2(P_2, P_2)$, lies on the diagonal $P = Q$. Its coordinates P_2 are the solutions of the equation $1 - P = P^m$, and lie in the interval $(0,1)$. They are approximately equal to $1 - \ln m/m$. The coordinates of the third stationary point $R_3(P_3, Q_3)$ are approximately equal to $P_3 \approx 1 - \left(\frac{1}{m}\right)^{\frac{s}{s-1}}$, $Q_3 \approx \left(\frac{1}{m}\right)^{\frac{1}{s-1}}$. There are two pairs of points which exchange places: $R_4(0,0)$, $R_5(1,1)$ and $R_6(P_6, 0)$, $R_7(1, Q_7)$, where $P_6 \approx 1 - \left(\frac{1}{m}\right)^{\frac{s}{s-1}}$ and $Q_7 \approx \left(\frac{1}{m}\right)^{\frac{1}{s-1}}$.

A curve through the points R_1, R_3, R_2 divides the PQ unit square (see Figure 36) into a lower left and an upper right portion (in fact we are interested in the more restricted area $0 \le P \le 1$, $0 \le Q \le P$). The point $S_n(P_n, Q_n)$ in one portion goes into the point $S_{n+1}(P_{n+1} = 1 - Q_n^s P_n^{m-s}$, $Q_{n+1} = 1 - P_n^m)$ in the other. Therefore the arrows in Figure 36 show the direction in which the point $S_n(P_n, Q_n)$ moves as the depth n is increased by 2. The stationary point R_1 is stable. The nearby values of the probability characteristics p and q are satisfactory. For such values $P_n \to 1$ and $Q_n \to 0$ as the depth n increases. The stationary points R_4 and R_5 are also stable. The points $S_n(P_n, Q_n)$ approach them when the initial values $P_0 = p$, $Q_0 = q$ are not in the satisfactory zone. The remaining stationary points and pairs are unstable.

Our reason for going into such detail in this simple probability model is not that it is interesting in itself. Other models that better represent the real situation in respect of algorithms for games involve more parameters and are more complex. In some cases, however, the behaviors of the characteristics of positions A with different parameters y, $\boldsymbol{q}$, and $\boldsymbol{r}$ tend to resemble each other as the depth n increases, and the characteristics begin to vary as they do in our current simple model, in which they are independent of $y(A)$, $\boldsymbol{q}(A)$ and $\boldsymbol{r}(A)$.

Now suppose that the evaluation function $f(A)$ can take on values between 0 and 1, but the relevant probabilities depend only on the turn to move col A and the true score sc(A):

$$\begin{aligned} p_{\pm}(z) &:= \boldsymbol{P}\left(f(A) \substack{\leq \\ <} z \mid \operatorname{col} A = 0,\ \operatorname{sc}(A) = 1 \right) \\ &= \boldsymbol{P}\left(f(A) \substack{> \\ \geq} 1 - z \mid \operatorname{col} A = 1,\ \operatorname{sc}(A) = 0 \right); \\ q_{\pm}(z) &:= \boldsymbol{P}\left(f(A) \substack{\leq \\ <} z \mid \operatorname{col} A = 0,\ \operatorname{sc}(A) = 0 \right) \\ &= \boldsymbol{P}\left(f(A) \substack{> \\ \geq} 1 - z \mid \operatorname{col} A = 1,\ \operatorname{sc}(A) = 1 \right). \end{aligned}$$

Then in essence little has changed. If the game $\mathfrak{A}$ is completely uniform, with m moves at non-terminal positions, and if there are exactly s winning moves at each non-terminal position won for its own color, the general recursion equations have the form

$$\begin{aligned} &P_{\pm}(1, z, n+1) \\ &\quad = \prod_{i=1}^{s} (1 - P_{\mp}(0, 1-z, n)) \prod_{i=s+1}^{m} (1 - P_{\mp}(1, 1-z, n)) \\ &\quad = (1 - P_{\mp}(0, 1-z, n))^{s} (1 - P_{\mp}(1, 1-z, n))^{m-s}, \\ &P_{\mp}(0, z, n+1) \\ &\quad = \prod_{i=1}^{m} (1 - P_{\mp}(1, 1-z, n)) = (1 - P_{\mp}(1, 1-z, n))^{m}. \end{aligned}$$

We introduce the notation:

$$P_n(z) := \begin{cases} 1 - P_{\pm}(1, z, n), & n \text{ even}; \\ 1 - P_{\mp}(1, 1-z, n), & n \text{ odd}; \end{cases}$$

$$Q_n(z) := \begin{cases} 1 - P_{\pm}(0, z, n), & n \text{ even}; \\ 1 - P_{\mp}(0, 1-z, n), & n \text{ odd}. \end{cases}$$

In this notation the recursion equations we have just written coincide with the recursion equations for our earlier model in which the evaluation

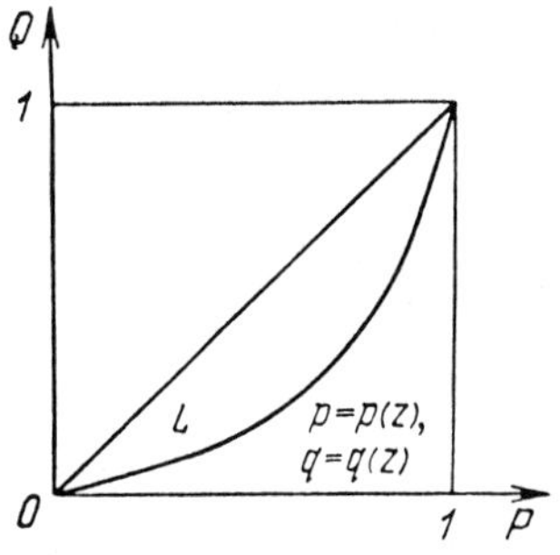

Figure 37

function had only the values 0 and 1:

$$1-P_{n+1}(z)=Q_n^s(z)P_n^{m-s}(z),$$
$$1-Q_{n+1}(z)=P_n^m(z),$$

with

$$P_0(z)=1-p_{\pm}(z),$$
$$Q_0(z)=1-q_{\pm}(z).$$

Suppose that the functions $p_{\pm}(z)$ and $q_{\pm}(z)$ are strictly monotone, continuous, and differentiable. Such functions can approximate arbitrary functions $p_{\pm}(z)=\boldsymbol{P}(f(A)\lesseqgtr z\,|\,\mathrm{col}\,A=0,\ \mathrm{sc}(A)=1)$ and $q_{\pm}(z)=\boldsymbol{P}(f(A)\lesseqgtr z\,|\,\mathrm{col}\,A=0,\ \mathrm{sc}(A)=0)$. Then $p_+(z)=p_-(z)$, $q_+(z)=q_-(z)$; in the interval $[0,1]$ these functions vary from 1 to 0. If the evaluation function $f(A)$ is relevant, the inequality $P_0(z)>Q_0(z)$ holds over the range $0<z<1$. The functions $P_0(z)$, $Q_0(z)$ define a segment of a curve, parameterized by z, lying in the unit square $0\le P\le 1$, $0\le Q\le 1$ of the PQ plane (see Figure 37). More exactly, the segment lies in the triangle $0\le P\le 1$, $0\le Q\le P$, joins the points $O(0,0)$ and $I(1,1)$, and is the graph of a monotone increasing continuous and differentiable function $Q=\varphi(P)$ (its derivative may equal $+\infty$ at some points).

The curves parameterized by the functions $P_n(z)$ and $Q_n(z)$ for $n=1,2,\ldots$ have similar properties. In fact, if $P_{n-1}(z)$ and $Q_{n-1}(z)$ are continuous, differentiable, and monotone decreasing (increasing), varying from 1 to 0 (0 to 1) in the interval $0\le z\le 1$, the functions $P_n(z)=1-Q_{n-1}^s(z)P_{n-1}^{m-s}(z)$ and $Q_n(z)=1-P_{n-1}^m(z)$ are also continuous and differentiable; they increase monotonely if $P_{n-1}(z)$ and $Q_{n-1}(z)$ decrease monotonely: they decrease monotonely if the latter functions increase monotonely. They vary from 0 to 1 or 1 to 0, as the case may be. Furthermore, for $0<z<1$

$$\begin{aligned}P_n(z)&=1-Q_{n-1}^s(z)P_{n-1}^{m-s}(z)\\&>1-P_{n-1}^s(z)P_{n-1}^{m-s}(z)=1-P_{n-1}^m(z)=Q_n(z).\end{aligned}$$

Now, for a model of depth n, and using Zermelo's formula for selecting a best move, we define the probability of selecting a winning move at a White position A whose score is 1, or at a Black position whose score is 0. Suppose that at the position A there are m moves $\Psi_i = (A, B_i)$ ($i = 1, 2, \ldots, m$), of which s win and the rest lose. With probability $|Q'_{n-1}(z)|dz$ the position B_i arising after the move $\Psi_i = (A, B_i)$ will have the model score $\operatorname{msc}_n(B_i) = z$ (this position is lost for its own color $\operatorname{col} B_i$). With probability $Q_{n-1}^{s-1}(z) P_{n-1}^{m-s}(z)$ the score z of the position B_i is the maximum of the scores of all the positions B_j ($j = 1, 2, \ldots, m$). (In all the positions B_j the move belongs to the color opposite to $\operatorname{col} A$, so that the scores msc_{n-1} (B_j) cannot be less favorable to the color $\operatorname{col} B_j = \operatorname{col} B_i$ than $\operatorname{msc}_{n-1}(B_i) = z$.) Thus the probability of choosing one of the s winning moves is equal to

$$\pi_n = s \int_0^1 Q_{n-1}^{s-1}(z) P_{n-1}^{m}(z) |Q'_{n-1}(z)| dz.$$

The functions $P_{n-1}(z)$ and $Q_{n-1}(z)$ define the curve $L_{n-1}\{P_{n-1}(z), Q_{n-1}(z)\}_0^1$, and

$$\pi_n = s \int_{L_{n-1}} Q_{n-1}^{s-1} P_{n-1}^{m-s} dQ = s(m-s) \int_{\Omega} Q_{n-1}^{s-1} P_{n-1}^{m-s-1} dP\, dQ,$$

where the curve L_{n-1} and the region Ω are as shown in Figure 38. If there is a point R_0 on the curve L_0 and lying in the regions of stability shown in Figures 36 and 38, there are points R_n on the curves L_n ($n = 1, 2, \ldots$) which tend to the point $R_\infty(1,0)$. The domains of integration Ω_n approach the unit square $0 \leq P \leq 1$, $0 \leq Q \leq 1$. Then

$$\lim_{n \to \infty} \pi_n = s(m-s) \int_0^1 \int_0^1 P^{m-s-1} Q^{s-1} dP dQ = 1.$$

If, however, the curve L_0 lies wholly outside the region of stability, the curves L_n converge to the diagonal of the unit square (a rigorous proof can be developed by using the theory of dynamic systems in a plane region, and is omitted here). Then

$$\lim_{n \to \infty} \pi_n = s(m-s) \int_0^1 \int_0^1 P^{m-s-1} Q^{s-1} dP dQ = s/m.$$

The values of $P_n(z)$ and $Q_n(z)$ are shown in Table 4 for several values of z together with the probabilities of selecting a winning move when $m = 10$, $s = 2$, $n = 0.2.4.6.8$, and the functions $P_0(z)$ and $Q_0(z)$ are equal respectively to $1 - z^4$ and $1 - z\sqrt{z}$, $1 - z^4$ and $1 - \sqrt{z}$, $1 - z^8$ and $\sqrt[8]{z}$. [V. P. Akimov studied the probability of choosing a winning move in the current model and carried out the computations present in this section.]

We now consider models in which the value of the evaluation function $f(B)$ [in the position B after the move $\Psi = (A, B)$ from the position A] depends on the value of $f(A)$ at the origin A of the move Ψ. For simplicity we shall suppose for the present that we have a completely uniform game $\mathfrak{A}_{m,N}$ with s winning moves for the side having color $\operatorname{col} A$ at the non-terminal position A, and with a search depth N much greater than the depth n of the models. It is comparatively easy to derive similar formulae

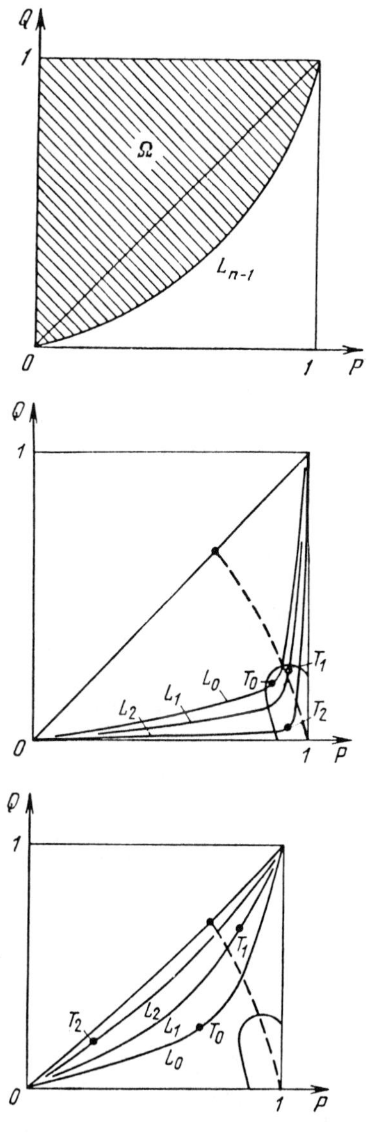

Figure 38

when we are given the distribution functions for the values of m and s, i.e. for the total number of moves and the number of winning moves at the positions of interest.

If the evaluation function $f(B)$ and the true score $\mathrm{sc}(B)$ take on only the values 0 and 1, and their probability characteristics depend only on 1) the color col B, 2) the score $\mathrm{sc}(B)$, 3) the score $\mathrm{sc}(A)$ of the antecedent position A from which the move $\Psi = (A.B)$ was made, and 4) the value of $f(A)$, and if they satisfy the symmetry conditions, they are determined by six

Table 4. Probabilities that Model Scores will be Correct for Shannon Models with Evaluation Functions Taking Intermediate Values

$m = 10, \quad s = 2$

$P_0(x) = x^4, \quad Q_0(x) = x\sqrt{x}$

Search depth n	$P_n(0,57)$	$Q_n(0,57)$	$P_n(0,58)$	$Q_n(0,58)$	R_n
0	0.1056	0.4303	0.1132	0.4414	0.4173
2	0.1444	0.2402	0.1770	0.2809	0.2618
4	0.1462	0.1631	0.2923	0.3158	0.2264
6	0.1082	0.1104	0.7385	0.4716	0.2259
8	0.0221	0.0222	0.9999	0.9999	0.2063

$P_0(x) = x^4 \quad Q_0(x) = \sqrt{x}$

Search depth n	$P_n(0,47)$	$Q_n(0,47)$	$P_n(0,48)$	$Q_n(0,48)$	R_n
0	0.0488	0.6856	0.0531	0.6928	0.7867
2	0.0895	0.5038	0.1068	0.5330	0.5137
4	0.1378	0.2906	0.2186	0.3966	0.2820
6	0.1572	0.1885	0.5529	0.5950	0.2144
8	0.1548	0.1594	0.9973	0.9974	0.2010

$P_0(x) = x^8 \quad Q_0(x) = \sqrt[8]{x}$

Search depth n	$P_n(0,42)$	$Q_n(0,42)$	$P_n(0,43)$	$Q_n(0,43)$	R_n
0	0.0000096	0.8972	0.00967	0.9301	0.9941
2	0.000008	0.9000	0.00827	0.9557	0.9982
4	0.00000007	0.9045	0.00626	0.9818	0.9998
6	0	0.9124	0.00370	0.9969	0.9999
8	0	0.9259	0.00132	0.9999	1.0000

Notation: m—Number of moves from any position, s—number of winning moves from a won non-terminal position, n—depth of the model, $P_n(x)$—probability that the model score $\text{msc}_n(A) \leq x$ for positions W in the original game, $Q_n(x)$—probability that the model score $\text{msc}_n(A) \leq x$ for position lost in the original game, R_n—probability of correctly choosing the best move.

parameters:

$$p(x_0, x, y) := \boldsymbol{P}(f(B) = |1 - \text{col } A| \,|\, \text{sc}(A) = |x_0 - \text{col } A|, \text{sc}(B) = |x_1 - \text{col } A|, f(A) = |y - \text{col } A|);$$

Thus

$$\begin{aligned}
p(1,1,1) &:= \boldsymbol{P}(f(B) = 1 | \text{col } A = 0, \text{sc}(A) = 1, \text{sc}(B) = 1, f(A) = 1)\\
&= \boldsymbol{P}(f(B) = 0 | \text{col } A = 1, \text{sc}(A) = 0, \text{sc}(B) = 0, f(A) = 0);\\
p(1,1,0) &:= \boldsymbol{P}(f(B) = 1 | \text{col } A = 0, \text{sc}(A) = 1, \text{sc}(B) = 1, f(A) = 0)\\
&= \boldsymbol{P}(f(B) = 0 | \text{col } A = 1, \text{sc}(A) = 0, \text{sc}(B) = 0, f(A) = 1);\\
p(1,0,1) &= \boldsymbol{P}(f(B) = 1 | \text{col } A = 0, \text{sc}(A) = 1, \text{sc}(B) = 0, f(A) = 1)\\
&= \boldsymbol{P}(f(B) = 0 | \text{col } A = 1, \text{sc}(A) = 0, \text{sc}(B) = 1, f(A) = 0);\\
p(1,0,0) &= \boldsymbol{P}(f(B) = 1 | \text{col } A = 0, \text{sc}(A) = 1, \text{sc}(B) = 0, f(A) = 0)\\
&= \boldsymbol{P}(f(B) = 0 | \text{col } A = 1, \text{sc}(A) = 0, \text{sc}(B) = 1, f(A) = 1);\\
p(0,0,1) &= \boldsymbol{P}(f(B) = 1 | \text{col } A = 0, \text{sc}(A) = 0, \text{sc}(B) = 0, f(A) = 1)\\
&= \boldsymbol{P}(f(B) = 0 | \text{col } A = 1, \text{sc}(A) = 1, \text{sc}(B) = 1, f(A) = 0);\\
p(0,0,0) &= \boldsymbol{P}(f(B) = 1 | \text{col } A = 0, \text{sc}(A) = 0, \text{sc}(B) = 0, f(A) = 0)\\
&= \boldsymbol{P}(f(B) = 0 | \text{col } A = 1, \text{sc}(A) = 1, \text{sc}(B) = 1, f(B) = 1).
\end{aligned}$$

The parameters $p(0,1, y)$ do not exist, since from a position that loses for its own color all moves are losing moves.

Let $P(x, y, n)$ be the probabilities of obtaining the model score $|1 - \text{col } A|$, given 1) the score $|x - \text{col } A|$ for the base position A, 2) the value $f(A) = |y - \text{col } A|$ for the evaluation function, and 3) the depth n of the model:

$$P(x, y, n) := \boldsymbol{P}(\text{msc}_n(A) = |1 - \text{col } A| \,|\, \text{sc}(A) = |x - \text{col } A|, f(A) = |y - \text{col } A|)$$

i.e.

$$\begin{aligned}
P(1,1,n) &:= \boldsymbol{P}(\text{msc}_n(A) = 1 | \text{col } A = 0, \text{sc}(A) = 1, f(A) = 1)\\
&= \boldsymbol{P}(\text{msc}_n(A) = 0 | \text{col } A = 1, \text{sc}(A) = 0, f(A) = 0);\\
P(1,0,n) &:= \boldsymbol{P}(\text{msc}_n(A) = 1 | \text{col } A = 0, \text{sc}(A) = 1, f(A) = 0)\\
&= \boldsymbol{P}(\text{msc}_n(A) = 0 | \text{col } A = 1, \text{sc}(A) = 0, f(A) = 1);\\
P(0,1,n) &:= \boldsymbol{P}(\text{msc}_n(A) = 1 | \text{col } A = 0, \text{sc}(A) = 0, f(A) = 1)\\
&= \boldsymbol{P}(\text{msc}_n(A) = 0 | \text{col } A = 1, \text{sc}(A) = 1, f(A) = 0);\\
P(0,0,n) &:= \boldsymbol{P}(\text{msc}_n(A) = 1 | \text{col } A = 0, \text{sc}(A) = 0, f(A) = 0)\\
&= \boldsymbol{P}(\text{msc}_n(A) = 0 | \text{col } A = 1, \text{sc}(A) = 1, f(A) = 1).
\end{aligned}$$

It is easily seen that

$$P(x,1,0)=1,$$
$$P(x,0,0)=0 \qquad (x=0,1).$$

The general recursion equations for the values of the characteristics $P(x, y, n+1)$ can be reduced to the form

$$1-P(1, y, n+1)$$
$$=(p(1,1, y)P(0,0, n)+(1-p(1,1, y))P(0,1, n))^{s}$$
$$\times(p(1,0, y)P(1,0, n)+(1-p(1,0, y))P(1,1, n))^{n-s},$$
$$1-P(0, y, n+1)$$
$$=(p(0,0, y)P(1,0, n)+(1-p(0,0, y))P(1,1, n))^{m} \quad (y=0,1).$$

A study of the way in which the probabilities $P(x, y, n)$ vary with an increase in n, as a function of the probability charactistics $p(x, y, n)$, is complicated by the large number (i.e. 6) of these. Therefore we do not have even a qualitative knowledge of the general situation. Table 5 lists the values of $P(x, y, n)$ for $n=0,1,2,3,4,5,6$ and for a few sets of values of $p(x_0, x_1, y)$. The computations were made for $m=10$, $s=2$.

If for some depth n_0 the value of $P(1,0, n_0)$ is close to that of $P(1,1, n_0)$ and the value of $P(0,1, n_0)$ is close to that of $P(0,0, n_0)$, then for $n>n_0$ our recursion equations imply that $P(1,1, n)$ and $P(1,0, n)$ will be close to P_n, and that $P(0,1, n)$ and $P(0,0, n)$ will be close to Q_n for models in which the values of the evaluation functions $f(B)$ at the positions B after the moves $\Psi=(A.B)$ are uncorrelated with the values of $f(A)$ at those preceding positions A for which $P_{n_0} \approx P(1,1, n_0) \approx P(1,0, n_0)$ and $Q_{n_0} \approx P(0,1, n_0) \approx P(1,1, n_{n_0})$. The probabilities $P(1,1, n)$ and $P(1,0, n)$ will tend to 1; $P(0,1, n)$ and $P(0,0, n)$ will tend to 0, whenever for some depth n_0 of the Shannon model tree the point in the PQ-plane with coordinates $P=P(1,0, n_0)$, $Q=P(0,1, n_0)$ lies in the region of stability depicted in Figure 36.

They will converge together, alternately to 1 and 0, if the point with coordinates $P=P(1,1, n_0)$, $Q=P(0,0, n_0)$ lies at some distance outside that region.

Let us pause to estimate some values of of the parameters $p(x_0, x_1, y)$ that specify the increase in reliability of the model scores with an increase in the depth of the model tree. Let $p(1,1,1)=p(1,0,1)=p(0,0,1)=1$; $p(1,1,0)=0.7$; $p(1,0,0)=p(0,0,0)=0.01$. The fact that the first three parameters are equal to 1 means that the evaluation function $f(A)$ is based on a highly optimistic estimate of the positions B resulting from arbitrary moves from A, at which the evaluation function favors the side having the move. An example would be a purely material-based evaluation function since one does not lose material on one's own move. The values $p(1,1,0) = 0.7$ and $p(0,0,0) = 0.1$ indicate that in approximately 90% of one's own positions that are evaluated as bad, the true opinion is found after a search of depth $n = 1$. However, our current model corresponds only very roughly to the situations that occur in real games, with the evaluation functions that are actually used.

We may consider an analogous model in which the evaluation function $f(A)$ may take on values intermediate between 0 and 1. We introduce the following notation:

$$P_{\pm}(x, y, z, n) :=$$
$$= \boldsymbol{P}\Big(\mathrm{msc}_n(A) \underset{>}{\geq} z | \mathrm{col}\, A = 0, \mathrm{sc}(A) = x, f(A) = y\Big)$$
$$= \boldsymbol{P}\Big(\mathrm{msc}_n(A) \underset{\leq}{<} 1 - z | \mathrm{col}\, A = 1, \mathrm{sc}(A) = 1 - x, f(A) = 1 - y\Big);$$

$$p(x_0, x_1, y_0, y_1) :=$$
$$= \boldsymbol{P}(f(B) \leq y_1 | \mathrm{col}\, A = 0, \Psi = (A, B) \in \mathfrak{A}, \mathrm{sc}(A) = x_0,$$
$$\mathrm{sc}(B) = x, f(A) = y_0)$$
$$= \boldsymbol{P}(f(B) \geq 1 - y_1 | \mathrm{col}\, A = 1, \Psi = (A, B) \in \mathfrak{A}, \mathrm{sc}(A) = 1 - x_0,$$
$$\mathrm{sc}(B) = 1 - x_1, f(A) = 1 - y_2);$$

In such a model the general recursion formula has the form:

$$1 - P_{\pm}(1, y, z, n+1)$$
$$= \left(\int_0^1 P_{\mp}(0, 1-\mu, 1-z, n)\, dp(1, 1, y, \mu)\right)^s$$
$$\times \left(\int_0^1 P_{\mp}(1, 1-\mu, 1-z, n)\, dp(1, 0, y, \mu)\right)^{m-s}.$$

$$1 - P_{\pm}(0, y, z, n+1)$$
$$= \left(\int_0^1 P_{\mp}(1, 1-\mu, 1-z, n)\, dp(0, 0, y, \mu)\right)^m$$

We use Stieltjes integration for the integrals entering the formula; then the formula is the same in the cases when the evaluation function $f(A)$ may take on all values in the interval $[0,1]$ or only a finite set of them.

Table 5. Probability of Correct Scores for a Shannon Model when the Value of the Evaluation Function at Points B Arising After Moves $\Psi = (A,B)$ Depends on the Value of the Evaluation Function at A Before the Move Ψ

$p(1,1,0) = 0.8 \quad p(1,0,0) = p(0,0,0) = 0.01$

n	$P(1,1,n)$	$P(1,0,n)$	$P(0,1,n)$	$P(0,0,n)$
0	1.0000	0.0000	1.0000	0.0000
1	1.0000	0.9631	1.0000	0.0956
2	0.9932	0.9238	0.3135	0.0037
3	1.0000	0.9959	0.5474	0.0732
4	0.9950	0.9720	0.0398	0.0005
5	1.0000	0.9999	0.2469	0.0515
6	0.9974	0.9918	0.0007	0.0000
7	1.0000	1.0000	0.0790	0.0267
8	0.9993	0.9986	0.0000	0.0000
9	1.0000	1.0000	0.0137	0.0072
10	0.9999	0.9999	0.0000	0.0000

	$p(1,1,0) = 0.6$, $p(1,0,0) = p(0,0,0) = 0.01$				$P(1,1,0) = 0.5.$ $p(1,0,0) = p(0,0,0) = 0.01$			
n	$P(1,1,n)$	$P(1,0,n)$	$P(0,1,n)$	$P(0,0,n)$	$P(1,1,n)$	$P(1,0,n)$	$P(0,1,n)$	$P(0,0,n)$
0	1.0000	0.0000	1.0000	0.0000	1.0000	0.0000	1.0000	0.0000
1	1.0000	0.8524	1.0000	0.0956	1.0000	0.7693	1.0000	0.0956
2	0.9975	0.7933	0.7976	0.0147	0.9989	0.7054	0.9274	0.0228
3	1.0000	0.8964	0.9013	0.0450	1.0000	0.7815	0.9695	0.0398
4	0.9992	0.8511	0.6650	0.0106	0.9998	0.7498	0.9150	0.0219
5	1.0000	0.9535	0.6912	0.0230	1.0000	0.7853	0.9439	0.0269
6	0.9997	0.8891	0.5305	0.0076	0.9999	0.7685	0.9109	0.0217
7	1.0000	0.9764	0.5694	0.0139	0.9999	0.7867	0.9281	0.0239
8	0.9999	0.9192	0.3786	0.0049	0.9999	0.7773	0.9092	0.0217

	$p(1,1,0) = 0.6$, $p(1,0,0) = p(0,0,0) = 0.01$				$p(1,1,0) = 0.5$, $p(1,0,0) = p(0,0,0) = 0.01$			
n	$P(1,1,n)$	$P(1,0,n)$	$P(0,1,n)$	$P0,0,n)$	$P(1,1,N)$	$P(1,0,n)$	$P(0,1,n)$	$P(0,0,n)$
9	1.0000	0.9925	0.4281	0.0061	0.9999	0.7874	0.9194	0.0229
10	0.9999	0.9457	0.2128	0.0025	0.9999	0.7819	0.9084	0.0217
16	1.0000	0.9979	0.0001	0.0000				
20					0.9999	0.7872	0.9081	0.0217

	$P(1,1,0)=0.8$, $P(1,0,0)=P(0,0,0)=0.015$				$P(1,1,0)=0.6$, $p(1,0,0)=P(0,0,0)=0.015$			
n	$P(1,1,n)$	$P(1,0,n)$	$P(0,1,n)$	$P(0,0,n)$	$P(1,1,n)$	$P(1,0,n)$	$P(0,1,n)$	$P(0,0,n)$
0	1.0000	0.0000	1.0000	0.0000	1.0000	0.0000	1.0000	0.0000
1	1.0000	0.9646	1.0000	0.1403	1.0000	0.8582	1.0000	0.1403
2	0.9853	0.9029	0.3029	0.0053	0.9942	0.7695	0.7832	0.0211
3	1.0000	0.9963	0.6398	0.1488	0.9999	0.9013	0.9271	0.0879
4	0.9785	0.9390	0.0363	0.0007	0.9966	0.8227	0.6462	0.0152
5	1.0000	0.9999	0.4668	0.2000	1.0000	0.9317	0.8579	0.0582
6	0.9600	0.9358	0.0005	0.0000	0.9981	0.8583	0.5070	0.0107
7	1.0000	1.0000	0.4851	0.3377	1.0000	0.9576	0.7831	0.0395
8	0.8859	0.8652	0.0000	0.0000	0.9989	0.8871	0.3514	0.0067
9	1.0000	1.0000	0.7650	0.7032	1.0000	0.9796	0.6982	0.0274
10	0.5056	0.4880	0.0000	0.0000	0.9994	0.9128	0.1866	0.0032
18					1.0000	0.9949	0.0000	0.0000

Notation: m—number of moves from non-terminal positions, s—number of winning moves from won non-terminal positions, n—depth of the model, $p(x, y, z)$—probability that $f(B)=|1-\text{col}\,A|$ provided that $\text{sc}(A)=|x-\text{col}\,A|$, $\text{sc}(B)=|y-\text{col}\,A|$, $F(A)=|z-\text{col}\,A|$, $P(x, z, u)$—probability that $\text{msc}(A)=|1-\text{col}\,A|$ provided that $\text{msc}(A)=|x-\text{col}\,A|$, $f(A)=|z-\text{col}\,A|$. In all the examples $m=10$, $s=2$, $p(1,1,1)=p(1,0,1)=p(0,0,1)=1$.

The calculations in the above formulae have a definite meaning, but for qualitative studies we may use models in which the evaluation function takes on only the two values 0 and 1. In such models we can reflect various relationships between the values of the evaluation function and the properties of positions. The characteristics of positions may depend not only on the value of the evaluation function but also on other parameters such as, for example, the so-called sharpness of the position. Later we shall offer three examples in order to show how to construct models of this kind; we shall examine in detail the influence of various parameters. We can construct combined models by similar methods, in which we take into account several parameters.

Let the number of winning moves at various positions depend on the value of the evaluation function $f(A)$. In the simplest case we suppose that in a winning position where $f(A)=1$ there are s_1 winning moves, and in a winning position where $f(A)=0$ there are only $s_2 < s_1$ such moves. If the value of $f(B)$ depends probabilistically only on the true scores of the positions A and B (where A is the position from which the move $\Psi=(A, B)$ leading to B was made) and the value of the evaluation function $f(A)$ at A, and if the parameters $p(x_0, x_1, y)$ have the same meaning as in the second model considered in this section, then the general recursion formula has the form:

$$
\begin{aligned}
1 - P(1,1,n+1)& \\
= \big(p(1,1,1)&P(0,0,n) \\
&+ (1-p(1,1,1))P(0,1,n)\big)^{s_1}\big(p(1,0,1)P(1,0,n) \\
&+ (1-p(1,0,1))P(1,1,n)\big)^{m-s_1}, \\
1 - P(1,0,n+1)& \\
= \big(p(1,1,0)&P(0,0,n) \\
&+ (1-p(1,1,0))P(0,1,n)\big)^{s_2}\big(p(1,0,0)P(1,0,n) \\
&+ (1-p(1,0,0))P(1,1,n)\big)^{m-s_2}, \\
1 - P(0,1,n+1)& \\
= \big(p(0,0,1)&P(1,0,n) + (1-p(0,0,1))P(1,1,n)\big)^{m}, \\
1 - P(0,0,n+1)& \\
= \big(p(0,0,0)&P(1,0,n) + (1-p(0,0,0))P(1,1,n)\big)^{m}.
\end{aligned}
$$

Suppose that the positions are either quiet or acute, and that to every position A there correspond three parameters on which the probability characteristics depend: the color $\mathrm{col}(A)$ of the side having the turn to move; the true score $\mathrm{sc}(A)$; and the acuteness $\varphi(A)$ of the position (for an acute position $\varphi(A)=1$, for a quiet position $\varphi(A)=0$). In an acute won position, for example, there will be only one winning move, which will lead with probability r_1 to an acute position and with probability $1-r_1$ to a quiet position. Both the acute and the quiet positions, naturally, are lost for the opponent. The remaining $m-1$ moves lead to positions won for the opponent, acute with probability r_2 and quiet with probability $1-r_2$. All moves from an acute lost position lead to positions won for the opponent, acute with probability r_3 and quiet with probability $1-r_3$. For simplicity we shall suppose that all moves from a quiet position lead to quiet positions and that from a won quiet position there are $s>1$ winning moves.

We denote by $P(1,n)$ the probability of correctly scoring an acute won position, and by $P(0,n)$ the probability of correctly scoring a quiet won position; $Q(1,n)$ and $Q(0,n)$ are the corresponding probabilities that acute and quiet lost positions will be evaluated as winning positions. In all the above definitions the parameter n represents the depth of a Shannon model for which the score of the base position is defined by the use of Zermelo's formula. The general recursion formula for such a model has the form:

$$
\begin{aligned}
1 - P(0,n+1) &= Q^s(0,n)P^{m-s}(0,n), \\
1 - Q(0,n+1) &= P^m(0,n), \\
1 - P(1,n+1) &= \big(r_1Q(1,n) + (1-r_1)Q(0,n)\big) \\
&\qquad \times \big(r_2P(1,n) + (1-r_2)P(0,n)\big)^{m-1}, \\
1 - Q(1,n+1) &= \big(r_3P(1,n) + (1-r_3)P(0,n)\big)^{m}.
\end{aligned}
$$

We must also prescribe the initial probabilities $P(0,0)=p_0$, $P(1,0)=p_1$, $Q(0,0)=q_0$, $Q(1,0)=q_1$.

Lastly we contemplate a model in which the total number of moves from a position, and the number of winning moves from a won position, may vary from position to position. For simplicity we shall again limit ourselves to the minimal number of parameters for the probability characteristics. Suppose these parameters are the color col A of the side having the move at A and the true score $sc(A)$. We consider the following characteristics: $\pi(m,s)$–the probability that from a won position we may make m moves, of which s win; $\rho(m)$–the probability that from a lost position A we may make m moves—all, naturally, losing. Thus $\pi(0,0)$ and $\rho(0)$ are the corresponding probabilities that won and lost positions will turn out to be terminal. The general recursion formula for our model has the form:

$$1-P_{n+1}=\pi(0,0)+\sum_{m=1}^{M}\sum_{s=1}^{m}\pi(m,s)Q_n^sP_n^{m-s},$$

$$1-Q_{n+1}=\sum_{m=1}^{M}\rho(m)P_n^m,$$

where M is the maximum number of moves that can be made from any position in our game, while P_n and Q_n have the same meaning that we assigned them in our first model. As before, we must prescribe the initial values $P_0=p$ and $Q_0=q$.

In conclusion we again recall that by similar methods we can construct models in which the probability characteristics depend on all the parameters we have discussed, including the number of admissible moves at the positions being studied, and on parameters not discussed in this section.

Construction of Models and Calculation of Model Scores on a Probabilistic Basis

In this section we consider some examples (due to N. E. Kosatcheva, who also carried out the necessary computations) for the case in which the probability of correctly assessing the score at the base position can be increased if we replace the classical Shannon model by one in which the branches in the search tree have varying lengths, or if we change the method for computing the scores at positions in the model. The basic methods for cutting down the number of positions to be examined in the search are founded on the following considerations:

(1) The search depth from a position B in the model game tree may depend on the value of the evaluation function $f(B)$ and perhaps on other parameters defined there.

(2) In determining the scores of positions in the model we may bypass Zermelo's formula and instead take account of the fact that several moves from a won position B should lead to a position with a value of $f(B)$ favorable to col B.
(3) The probability that a move from a given position will turn out to be the best one may be used not only to define the order in which the moves should be examined but also to determine the search depth.

All these opportunities will be illustrated for some probabilistic assumptions that are as simple as possible, in order to see what results they lead to. Simple as these assumptions are, they are applicable in more complex cases and the methods we illustrate may be combined.

Let B be the position arising after the move $\Psi = (A, B)$ and suppose that $f(B)$ is correlated with the true scores of A and B and with the value of $f(A)$. Then we are given the values of the probability characteristics $p_0(x_0, x_1, y)$ defined in the preceding section. With our standard assumptions (m moves from any non-terminal position, and s winning moves from a won position) for a Shannon model in which all nodes are inspected to a prescribed depth n, we have the recursion equations

$$\begin{aligned} 1 - P(1, y, n+1) &= \big(p(1,1,y)P(0,0,n) + (1-p(1,1,y))P(0,1,n)\big)^s \big(p(1,0,y) \times P(1,0,n) \\ &\quad + (1-p(1,0,y))P(1,1,n)\big)^{m-2}. \end{aligned}$$

$$1 - P(0, y, n+1) = \big(p(0,0,y)P(1,0,n) + (1-p(0,0,y))P(1,1,n)\big)^m$$

where $P(x,0,1) = 1$ and $P(x,0,0) = 0$ $(x = 0,1)$.

As an example we list the values of $P(x, y, n)$ in Table 6 for $n = 0.6$ and for $p(1,1,1) = p(1,0,1) = p(0,1,1) = 1$; $p(1,1,0 = 0.8$; $p(1,0,0) = p(0,0,0) = 0.02$; $m = 10$, $s = 2$. Inspection of the table shows that even for values of

Table 6. Probabilities of $P(x, y, n)$ as a Function of the Search Depth n

	Shannon Model $\mathfrak{A}_{SH,n}$				Model $\mathfrak{A}_{NK,n}$			
n	$P(1,1,n)$	$P(1,0,n)$	$P(0,1,n)$	$P(0,0,n)$	$P(1,1,n)$	$P(1,0,n)$	$P(0,1,n)$	$P(0,0,n)$
0	1.0000	1.0000	1.0000	0.0000				
1	1.0000	0.9660	1.0000	0.1829				
2	0.9746	0.8807	0.2927	0.0068	1.0000	0.8807	1.0000	0.0068
3	1.0000	0.9967	0.7193	0.2414	1.0000	0.9586	0.7193	0.0236
4	0.9433	0.8865	0.0323	0.0008	0.9996	0.9737	0.3488	0.0084
5	1.0000	1.0000	0.7001	0.4491	0.9999	0.9943	0.2340	0.0091
6	0.7983	0.7507	0.0003	0.0000	0.9999	0.9971	0.0554	0.0017

n_0 as low as 2 the point $(P(1,1,n_0), P(0,0,n_0))$ lies outside of the region of stability depicted in Figure 36. Therefore the probabilities $P(x, y, n)$ do not tend to desirable limits but instead tend alternately to the limits 0 and 1. Note that the applicability of the above recursion equations is not limited to the classical Shannon model.

Suppose that we have constructed a model in which the search depths may vary from node to node and that we have specified the probabilities $P(x, y)$ (at the base position A with $\mathrm{sc}(A) = |x - \mathrm{col}\, A|$) that the value of the model evaluation function $f(A) = |y - \mathrm{col}\, A|$ will be $\mathrm{msc}(A) = |1 - \mathrm{col}\, A|$. (We assume that in our model the natural symmetry conditions for probabilities are satisfied.) We now set up a mapping between our model game $\mathfrak{A}_\mu$ and a value n_0 of the formal search depth n; e.g. the maximum depth of a branch in the search tree $\mathfrak{A}_{\mu, n_0} = \mathfrak{A}_\mu$. Then for $n > n_0$ the model games $\mathfrak{A}_{\mu, n}$ are defined as follows: their trees consist of all the nodes A_k in the original tree that have rank $r(A_k) = k \leq n - n_0$, the moves leading from them, and the subtrees $\mathfrak{A}_{\mu, n_0}(A_{n-n_0})$. The base positions of these trees have rank $r(A_{n-n_0}) = n - n_0$ and the subtrees themselves, by construction, are trees in the model game $\mathfrak{A}_\mu$. It is easily seen that for $n \geq n_0$ we have the following recursion equations, similar to those developed above:

$$\begin{aligned} 1 - P_\mu(1, y, n+1) &= \big(p(1,1,y)P_\mu(0,0,n) + (1 - p(1,1,y))P_\mu(0,1,n)\big)^s \\ &\quad \times \big(p(1,0,y)P_\mu(1,0,n) + (1 - p(1,0,y))P_\mu(1,1,n)\big)^{m-s}, \\ 1 - P_\mu(0, y, n+1) &= \big(p(0,0,y)P_\mu(1,0,n) + (1 - p(0,0,y))P_\mu(1,1,n)\big)^m. \end{aligned}$$

Let us consider a model game $\mathfrak{A}_{NK,2}$, with the following convention: if at the base position A the evaluation function has the value $f(A) = |1 - \mathrm{col}\, A|$ we take A as terminal with the model score $\mathrm{msc}_{NK,2} = f(A)$, else $\mathfrak{A}_{NK,2}(A)$ is taken as the tree of a Shannon model game $\mathfrak{A}_{SH,2}(A)$ of depth 2. This tree contains the position A, all moves from it, all positions B_i ($i = 1,2,\ldots, m$) to which these moves lead, and all moves in the game $\mathfrak{A}$ from the B_i. The positions reached by these latter moves are taken as terminal. Then

$$P_{NK}(1,1,2) = P_{NK}(0,1,2) = 1,$$
$$P_{NK}(x,0,2) = P_{SH}(x,0,2) \ (x = 0,1).$$

The values of the $P_{NK}(x, y, n)$ are shown in Table 6 for the displayed values of the parameters $m, s, p(x_0, x_1, y)$ and for $n = 2,3,4,5,6$. For $n = 6$ the point $(P_{NK}(1,0,n), P_{NK}(0,1,n))$ lies in the region of stability illustrated in Figure 36. This means that with further increase in the search depth n the probabilities $P_{NK}(1,1,n)$ and $P_{NK}(1,0,n)$ will tend to 1 while $P_{NK}(0,1,n)$ and $P_{NK}(0,0,n)$ will tend to 0.

The rules for constructing the model $\mathfrak{A}_{NK,n}$ may be so formulated that they appear as well-defined considerations with a sensible meaning. If the search depth n, from the base position A_0, is odd we must compute the evaluation function $f(A_{n-2})$ for the positions A_{n-2} of rank $n-2$. If the value is bad for us (the side with color col A_0), we should believe it and not consider moves from the positions A_{n-2}. If the depth is even we must make an auxiliary search of depth 2 in order to sharpen our earlier good opinion of the position A_{n-2}. Such a reasonable argument, however, fails to explain why for even n we should trust our earlier good opinion of the positions of rank $n-2$ but sharpen a bad opinion.

Let us now consider some possibilities arising from our rejection of Zermelo's formula as the means of defining the model scores of non-terminal positions. Some authors (e.g. [27]) propose that the score of a position A should be defined as the mean value of the most favorable (for col A) of the model scores at the positions B_i that can be reached in one move from A:

$$\operatorname{msc}(A) := \begin{cases} \max\limits_{i_1, i_2, \ldots, i_k} \dfrac{1}{k} \sum\limits_{h=1}^{k} \operatorname{msc}(B_{i_h}), \\ \text{White to move} \\ \min\limits_{i_1, i_2, \ldots, i_l} \dfrac{1}{l} \sum\limits_{h=1}^{l} \operatorname{msc}(B_{i_h}), \\ \text{Black to move} \end{cases}$$

When the numbers k and l are given, this method is known as the k, l-heuristic. However, the values of k and l may vary for positions of different ranks in the tree of the model game, and may also depend on other circumstances.

We can broaden the class of such formulae by considering weighted averages, writing

$$\operatorname{msc}_n(A) := \begin{cases} \varkappa_{w,0} f(A) + \sum\limits_{i=1}^{m} \varkappa_{wi} \operatorname{msc}(B_i), \\ \text{White to move;} \\ \varkappa_{b,0} f(A) + \sum\limits_{i=1}^{m} \varkappa_{bi} \operatorname{msc}(B_i), \\ \text{Black to move;} \end{cases}$$

where the positions B_i that arise after the moves $\Psi_i = (A, B_i)$ ($i = 1, 2, , \ldots, m$) are ordered by decreasing values of the model score $\operatorname{msc}(B_i)$ when A is a White position and in the reverse order when A is a Black position. The formulae we have proposed form a particular case. We have for them

$$\varkappa_{w,0} = \varkappa_{b,0} = 0, \quad \varkappa_{w,1} = \varkappa_{w,2} = \cdots = \varkappa_{w,k} = \frac{1}{k},$$

$$\varkappa_{w,k+1} = \cdots = \varkappa_{w,m} = 0, \quad \varkappa_{b,1} = \varkappa_{b,2} = \cdots = \varkappa_{b,l} = \frac{1}{l},$$

$$\varkappa_{b,l+1} = \cdots = \varkappa_{b,m} = 0.$$

In applying such formulae we should use a pruning rule weaker than the α, β-heuristic, since we cannot conclude that we have obtained at a position A a model score favorable to col A merely because we have found a score unfavorable to col B at a position B reached by a single move from A. The consequent increase in the number of positions to be inspected may be offset by a decrease in the search depth n that suffices to obtain values near 1 for the probabilities P_n and $1 - Q_n$ of obtaining correct model scores for the positions won and lost, respectively, for col A. The probability characteristics in the game $\mathfrak{A}$ may be such that the use of Zermelo's formula does not guarantee the convergence of P_n to 1 and of Q_n to 0, but the use of a more general formula does so.

The model scores defined by our formulae take on values in the open interval $(0,1)$ even in the case when the evaluation function $f(A)$ and the true score $\mathrm{sc}(A)$ take on only the values 0 and 1. Therefore the probabilities of various values of the model scores defined by our formulae are computed by equations containing integrals. We shall present such formulae for a completely uniform game $\mathrm{A}_{m,N}$ with s winning moves from every position won for its own side, and values of the evaluation function $f(A)$ that are independently distributed random variables depending probabilistically only on $\mathrm{sc}(A)$ and col A.

Suppose that we have a Shannon model with depth $n \ll N$ and that for positions of rank k we have the probabilities

$$P_{\pm n,k}(z) := \boldsymbol{P}\Big(|\mathrm{col}\,A - \mathrm{msc}(A)| \gtrless z | \mathrm{sc}(A) = |1 - \mathrm{col}\,A|, r(A) = k\Big);$$

$$Q_{\pm n,k}(z) := \boldsymbol{P}\Big(|\mathrm{col}\,A - \mathrm{msc}(A)| \gtrless z | \mathrm{sc}(A) = \mathrm{col}\,A, r(A) = k\Big);$$

The values of the $P_{\pm n,n}(z)$ amd $Q_{\pm n,n}(z)$ are determined by the probability characteristics of the evaluation function $f(A)$:

$$P_{\pm n,n}(z) := \boldsymbol{P}\Big(|\mathrm{col}\,A - f(A)| \gtrless z | \mathrm{sc}(A) = |1 - \mathrm{col}\,A|, r(A) = n\Big);$$

$$Q_{\pm n,n}(z) := \boldsymbol{P}\Big(|\mathrm{col}\,A - f(A)| \gtrless z | \mathrm{sc}(A) = \mathrm{col}\,A, r(A) = n\Big).$$

The recursion equations for these probabilities have the form

$$1-P_{\pm n,k-1} = \int_{\Omega_{\pm}(z)} \left| \prod_{i=1}^{s} dQ_{n,k}(\zeta_i) \prod_{i=s+1}^{m} dP_{n,k}(\zeta_i)\, dp_{k-1}(\zeta_0) \right|,$$

$$1-Q_{\pm n,k-1} = \int_{\Omega_{\pm}(z)} \left| \prod_{i=1}^{m} dP_{n,k}(\zeta_i)\, dq_{k-1}(\zeta_0) \right|,$$

where $\Omega_{\pm}(z)$ is the region in the $(m+1)$-dimensional unit cube $(0 \le \zeta_i \le 1)_0^m$ whose points satisfy the condition

$$\varkappa_0\zeta_0 + \sum_{h=1}^{m} \varkappa_h(1-\zeta i_h) \genfrac{}{}{0pt}{}{<}{\le} z$$

for an arbitrary permutation $\begin{pmatrix} 1_2 \ldots m \\ i_1 i_2 \ldots i_m \end{pmatrix}$ (it suffices that the condition be satisfied for those permutations that carry the sequence of coordinates $(\zeta_1, \zeta_2, \ldots, \zeta_m)$ into a monotone non-increasing sequence). The set functions $p_k(z)$ and $q_k(z)$ in the recursion equations represent the probabilities of the values of the evaluation function $f(A)$:

$$p_k(z) := \boldsymbol{P}(|\mathrm{col}\, A - f(A)| \ge z | \mathrm{sc}(A) = |1 - \mathrm{col}\, A|,\ r(A) = k):$$

$$q_k(z) := \boldsymbol{P}(|\mathrm{col}\, A - f(A)| \ge z | \mathrm{sc}(A) = \mathrm{col}\, A,\ r(A) = k).$$

Thus the probability characteristics of the evaluation function $f(A)$ may depend on the rank $r(A)$ of the position A.

The above formulae are complex, even under simple probabilistic postulates, and no investigation of them has been made. Some simpler methods were considered (for study, not for application) in the case of non-minimax search, i.e. a search in which Zermelo's method is not used in finding the scores of non-terminal positions. This investigation showed that in principle it is possible to obtain results better than those yielded by a minimax search.

Suppose that the model score $\mathrm{msc}(A)$ of a position A which is non-terminal in the model has the value $|1-\mathrm{col}\, A|$ when l or more of the moves $\Psi = (\mathrm{A}, \mathrm{B})$ leading from it lead to positions B with $\mathrm{msc}(B) = |1 - \mathrm{col} A|$, and the remaining moves lead to positions B with $\mathrm{msc}\, B = \mathrm{col}\, A$. Then for a position A of rank $r(A) = k-1$, the respective probabilities $P_{n,k-1}$ and $Q_{n,k-1}$ that $\mathrm{msc}(A) = |1 - \mathrm{col}\, A|$ when A is won or lost for col A satisfy the recursion equations

$$1-P_{n,k-1} = \sum_{g=0}^{l} \sum_{h=0}^{\min(s,l-g)} C_s^h (1-Q_{n,k})^h Q_{n,k}^{s-h} C_{m-s}^g (1-P_{n,k})^g P_{n,k}^{m-s-g}$$

$$1-Q_{n,k-1} = \sum_{g=0}^{l} C_m^g (1-P_{n,k})^g P_{n,k}^{m-g}$$

(we assume $l \le m-s$). The model scores at positions of rank k are equal to the corresponding values of the evaluation function. Accordingly,

$$P_{n,n} = p := \boldsymbol{P}(f(A) = |1-\operatorname{col} A| \,|\, \operatorname{sc}(A) = |1-\operatorname{col} A|);$$
$$Q_{n,n} = q := \boldsymbol{P}(f(A) = |1-\operatorname{col} A| \,|\, \operatorname{sc}(A) = \operatorname{col} A);$$

Later we shall consider the possibility of varying the threshold l for positions (l is a number of moves, not rank) in the tree of a Shannon model having several ranks. Let $P_n(l)$ and $Q_n(l)$ be the probabilities (for won and lost positions, respectively) that $\operatorname{msc}(A) = |1-\operatorname{col} A|$. When l is constant, it follows from our formulae that these probabilities satisfy the recursion equations

$$\begin{aligned}1 - P_{n+1}(l) &\\ = \sum_{g=0}^{l} \sum_{h=0}^{\min(s,l-g)} & C_s^h(1-Q_n(l))^h Q_n^{s-h}(l) C_{m-s}^g (1-P_n(l))^g P_n^{m-s-g}(l),\end{aligned}$$

$$1 - Q_{n+1}(l) = \sum_{g=0}^{l} C_m^g (1-P_n(l))^g P_n^{m-g}(l).$$

If we set $l=0$ we obtain our earlier recursion equations for a minimax search. The values of $P_0(l) = p$ and $Q_0(l) = q$ are found by the methods developed earlier. They do not depend on l.

Let $P_n(l) = 1-\varepsilon$, $Q_n(l) = \delta$. Then

$$\begin{aligned}\varepsilon' &= 1 - P_{n-1}\\ &= \sum_{g=0}^{l} C_{m-s}^g \varepsilon^g (1-\varepsilon)^{m-s-g} \left(\sum_{h=0}^{\min(s,l-g)} C_s^h (1-\delta)^h \delta^{s-h} \right)\\ &= C_s^{\min(l,s)} \Pi'_{m,s,l}(\varepsilon,\delta) \delta^{\max(0,s-l)},\\ \delta' &= Q_{n+1}\\ &= 1 - \sum_{g=0}^{l} C_m^g \varepsilon^g (1-\varepsilon)^{m-g} = \sum_{g=l+1}^{m} C_m^g \varepsilon^g (1-\varepsilon)^{m-g} = C_m^{l+1} \Pi''_{m,l}(\varepsilon) \varepsilon^{l+1},\\ \varepsilon'' &= 1 - P_{n+2}\\ &= C_s^{\min(l,s)} \Pi'_{m,s,l}(\varepsilon',\delta') \delta'^{\max(0,s-l)} = C_s^{\min(l,s)} \left(C_m^{l+1}\right)^{\max(0,s-l)}\\ &\qquad \times \Pi'_{m,s,l}(\varepsilon',\delta') \Pi''^{\max(0,s-l)}_{m,l}(\varepsilon) \varepsilon^{(l+1)\max(0,s-l)},\\ \delta'' &= Q_{n+2}\\ &= C_m^{l+1} \Pi''_{m,l}(\varepsilon') \varepsilon'^{l+1} = C_m^{l+1} \left(C_s^{\min(l,s)}\right)^{l+1} \Pi''_{m,l}(\varepsilon')\\ &\qquad \times \Pi'^{l+1}_{m,s,l}(\varepsilon,\delta) \delta^{(l+1)\max(0,s-l)},\end{aligned}$$

where the polynomial $\Pi'_{m,s,l}(\varepsilon,\delta)$ is equal to 1 for $\varepsilon = \delta = 0$, $\Pi''_{m,l}(\varepsilon) = 1$ for $\varepsilon = 0$, and

$$0 \le \Pi'_{m,s,l}(\varepsilon,\delta) < 1,$$
$$0 \le \Pi''_{m,l}(\varepsilon) < 1 \quad (0 < \varepsilon \le 1, 0 < \delta \le 1).$$

When $l < s$ these formulae yield

$$\varepsilon'' = C_s^l \left(C_m^{l+1}\right)^{s-l} \Pi'_{m,s,l}(\varepsilon', \delta') \Pi''^{s-l}_{m,l}(\varepsilon) \varepsilon^{(l+1)(s-l)},$$

$$\delta'' = C_m^{l+1} \left(C_s^l\right)^{s+1} \Pi''_{m,l}(\varepsilon') \Pi'^{l+1}_{m,s,l}(\varepsilon, \delta) \delta^{(l+1)(s-l)},$$

while for $l \geq s$

$$\varepsilon'' = \Pi'_{m,s,l}(\varepsilon', \delta'),$$

$$\delta'' = C_{ml+1} \Pi''_{m,l}(\varepsilon') \Pi'^{(l+1)}_{m,s,l}(\varepsilon, \delta).$$

Thus if $l \geq s$, P_n does not tend to 1, nor Q_n to 0, for any value of p or q. In real games the number s of winning moves at won terminal positions is not constant; therefore special significance attaches to positions from which there is only one winning move. In our model a non-minimax search using the same formulae at all positions in the search tree is of doubtful effectiveness. Also, all the scores in this section are based on deliberately simplified probabilistic assumptions about the games considered. We are interested in new approaches to the construction of models of programmable games, definition of model scores, and selection of moves. The application of such approaches requires supplementary work on more realistic probabilistic hypotheses and on experimental verifications.

We return to the question of which values of l promote rapid convergence of the $P_n(l)$ and $Q_n(l)$ to the values we desire. From the formulae derived above it follows that when ε and δ are small enough the rate of convergence is determined by their exponent $(l+1)(s-l)$, which attains its maximum when $l = \ulcorner(s-1)/2\urcorner$ (for even s the exponents have the same value for $l = \ulcorner(s-1)/2\urcorner$ and $l = \urcorner(s-1)/2\ulcorner$, but the coefficients of the $\delta^{(l-1)(s-l)}$ increase as l increases). The rate of convergence must be defined for other probabilistic assumptions also. Moreover, one must take into account the fact that the number of positions in the search tree increases with increasing l. Thus it may be useless to decrease the depth of the search tree by going to a non- minimax search method.

Some values of $P_0(l) = p$ and $Q_0(l) = q$ that lie outside the region of stability depicted in Figure 36 will lie within it for a non-minimax search and some value of $l > 0$. For example, let $m = 10$, $s = 3$, $p = 0.95$, and $q = 0.1$; Table 7 shows the results of the computations for $l = 0, 1$, $n = 0, 1, 2, 3, 4$. If $P_{n+2}(l) < P_n(l)$ and $Q_{n+2}(l) < Q_n(l)$, our recursion equations imply the inequalities $P_{n+3}(l) > P_{n+1}(l)$, $Q_{n+3}(l) > Q_{n+1}(l)$. Conversely, if $P_{n+2}(l) > P_n(l)$ and $Q_{n+2}(l) > Q_n(l)$, it follows that $P_{n+3}(l) < P_{n+1}(l)$ and $Q_{n+3}(l) < Q_{n+1}(l)$.

Accordingly, whenever for some $n = n_0$ the probabilities $P_{n_0+2}(l)$ and $Q_{n_0+2}(l)$ differ from $P_{n_0}(l)$ and $Q_{n0}(l)$ in the same direction, $P_n(l)$ and $Q_n(l)$ will not tend to the desired limits. For instance, if $P_{n_0+2}(l) > P_{n_0}(l)$ and $Q_{n_0+2}(l) > Q_{n_0}(l)$, for $n = n_0 + 2k$ these probabilities will increase monotonely, and for $n = n_0 + 2k + 1$ they will decrease monotonely. The

Table 7. Various Methods for Computing Model Scores

$m = 10, \quad s = 3$				
	$l = 0$		$l = 1$	
n	P_n	Q_n	P_n	Q_n
0	0.9500	0.1000	0.9500	0.1000
1	0.9993	0.4013	0.9802	0.0861
2	0.9357	0.0070	0.9817	0.0159
3	1.0000	0.4855	0.9993	0.0137
4	0.8856	0.0000	0.9994	0.0000

Notation: m—number of moves from non-terminal position, s—number of winning moves from a won non-terminal position, l—number of moves from a position A to a position B with $\mathrm{msc}(B) = |1 - \mathrm{col}\, A|$ when $\mathrm{msc}(A) = |1 - \mathrm{col}\, A|$, n—search depth, P_n—probability that $\mathrm{msc}_n(A) = |1 - \mathrm{col}\, A|$ when $\mathrm{sc}(A) = |1 - \mathrm{col}\, A|$, Q_n—probability that $\mathrm{msc}_n(A) = |1 - \mathrm{col}\, A|$ when $\mathrm{sc}(A) = \mathrm{col}\, A$.

probabilities $P_n(0)$ and $Q_n(0)$ behave in this way in our example, as shown in Table 7.

The equations determining the distances of $P_{n+2}(l)$ and $Q_{n+2}(l)$ from the desired limits imply the inequalities

$$\varepsilon'' \leq 6075\varepsilon^4,$$

$$\delta'' \leq 40\,5\delta^4$$

for $m = 10$, $s = 3$, $l = 1$ (equality holds only when the corresponding distance ε or δ is equal to zero). Thus if $0 \leq \varepsilon \leq \sqrt[3]{1/6075} \approx 0.054\ 8$ and $0 \leq \delta \leq \sqrt[3]{1/405} \approx 0.1352$ the probabilities $P_n(l)$ and $Q_n(l)$ will converge as desired. Our values of $P_0(l) = p = 1 - \varepsilon = 0.95$ and $Q_0(l) = \delta = q = 0.1$ lie in this region of convergence.

When $s = 2$, a minimax search corresponding to $l = 0$ yields the fastest convergence for the P_n and Q_n. Moreover, the region of convergence for $l = 1$ lies inside that for $l = 0$, depicted in Figure 36. For $l \geq 2$ the probabilities $P_n(l)$ and $Q_n(l)$ do not in general converge to the desired limits. Nevertheless, for some values of p and q, employing the equations for a non-minimax search in finding the model scores at positions of high rank may turn out to be useful. For instance, set $m = 10$, $s = 2$, $P_{n,n} = p = 0.95$, $Q_{n,n} = 0.005$. The values p and q lie outside the region of convergence depicted in Figure 36. There is no convergence even when we apply a non-minimax search with $l = 1$.

However, if we use the equations for a non-minimax search with $l = 1$ to find the model scores for positions of rank $n - 1$ the probabilities $P_{n,n-1}$ and $Q_{n,n-1}$ will have the respective values 0.9937 and 0.0862. The point $(P_{n,n-1}, Q_{n,n-1})$ lies in our region of convergence for a minimax search, so that the model scores $\operatorname{msc}(A_k)$ for positions A of rank $k < n-1$ may be determined by Zermelo's formula, and as k decreases they tend to their desired values.

The case $p = 0.995$, $q = 0.2$ requires a somewhat more complex treatment. The point (p, q) of the PQ- plane lies outside of the region of convergence depicted in Figure 36. Suppose we are dealing with a Shannon model of depth n. If we apply Zermelo's formula to determine the model scores for positions of rank $n-1$, we find that $P_{n,n-1} = 0.9600$, $Q_{n,n-1} = 0.0050$. If we apply a non- minimax method with $l = 1$ to the positions of rank $n-2$ we find $P_{n,n-2} = 0.9928$, $Q_{n,n-2} = 0.0581$.

The standard Zermelo formula should be applied to determine the model scores for positions of lower rank.

In these last two examples an even better result can be obtained by drawing into the determination of the model score at a position A not only the model scores $\operatorname{msc}(B)$ at the positions B reached by the moves $\Psi = (A, B) \in \mathfrak{A}$ but also the value of the evaluation function $f(A)$. Let us look at such a rule:

(1) if two or more moves lead from A to positions B with $\operatorname{msc}(B) = |1 - \operatorname{col} A|$, then $\operatorname{msc}(A) = |1 - \operatorname{col} A|$;
(2) if there are no such moves, then $\operatorname{msc}(A) = \operatorname{col} A$;
(3) if one and only one move leads to a position B with $\operatorname{msc}(B) = |1 - \operatorname{col} A|$ then $\operatorname{msc}(A) = f(A)$.

In a Shannon model with the condition that the model scores of all positions of higher rank are defined in the same way, the use of the above rules in computing the model scores for positions of rank k leads to the following recursion equations:

$$1 - P_{n,k} = Q^s_{n,k+1} P^{m-s}_{n,k+1} + \big(s Q^{s-1}_{n,k+1}(1 - Q_{n,k+1}) \times P^{m-s}_{n,k+1} + (m-s) Q^s_{n,k+1} P^{m-s-1}_{n,k+1}(1 - P_{n,k+1})\big)(1-p),$$

$$1 - Q_{n,k} = P^m_{n,k+1} + m P^{m-1}_{n,k+1}(1 - P_{n,k+1})(1-q).$$

Since normally $(1-p)$ is small and $(1-q)$ more or less near to 1 the probability $P_{n,k+1}$ is near to the value obtained by a minimax search, and $Q_{n,k+1}$ is near the value obtained by our earlier method with $l = 1$. If we apply the above rules to determine the model scores of all positions, the region of convergence of the P_n and Q_n to their desired values contains a portion lying outside the region depicted in Figure 36 for a minimax search. For example, with $m = 10$, $s = 2$ the point $p = 0.95$, $q = 0.15$ lies in the region of convergence.

We now consider probabilistic postulates relating to the ordering of moves in our game with respect to their expected quality. As before, we suppose that White and Black move alternately, that the true scores $\operatorname{sc}(A)$ and the evaluation function $f(A)$ take on the values 0 and 1, and that the model scores $\operatorname{msc}(A)$ of positions A in the tree of the model game $\mathfrak{A}_\mu$ are defined by the customary minimax search using the the values $f(F)$ of the evaluation function $f(F)$ of the terminal positions F in the model game $\mathfrak{A}_\mu$ (if F is also terminal in the original game $\mathfrak{A}$, then naturally $f(F) = \text{sc}(F)$). We stipulate that the following probabilistic postulates are satisfied:

(1) The conditional probabilities $p(F)$ and $q(F)$ that $f(F) = |1 - \text{col}\, F|$ for terminal positions F in the model game $\mathfrak{A}_\mu$, under the respective hypotheses that $\text{sc}(F) = |1 - \text{col}\, F|$ and that $\text{sc}(F) = \text{col}\, F$, are mutually independent.
(2) Also mutually independent for non- terminal positions $A \in \mathfrak{A}_\mu$ are the probabilities $\pi(A, \Omega)$ that if $\text{sc}(A) = |1 - \text{col}\, A|$ then the moves $\Psi = (A, B) \in \Omega \subset M(A)$, and only those moves, are winning moves.

$$\pi(A,\Omega) := \boldsymbol{P}\left(\text{sc}(B) = \begin{cases} |1-\text{col}\, A|, & \text{if } \Psi \in \Omega \\ \text{col}\, A, & \text{if } \Psi \notin \Omega \end{cases} \Big| \text{sc}(A) = |1-\text{col}\, A|\right).$$

For non-terminal positions A in the model game $\mathfrak{A}_\mu$, provided the subtrees $\mathfrak{A}_\mu(A)$ do not intersect, these postulates imply the mutual independence of the conditional probabilities $P(A)$ and $Q(A)$ that the model scores will have the value $\text{msc}(A) = |1 - \text{col}\, A|$ under the respective hypotheses that $\text{sc}(A) = |1 - \text{col}\, A|$ and that $\text{sc}(A) = \text{col}\, A$. Then the probabilities in question are given by:

$$P(A) = 1 - \sum_{\Omega \in M(A)} \pi(A,\Omega) \prod_{\Psi = (A,B) \in \Omega} Q(B) \prod_{\Psi = (A,B) \in M(A) \setminus \Omega} P(B),$$

$$Q(A) = 1 - \prod_{\Psi = (A,B) \in M(A)} P(B).$$

We may use dynamic programming in constructing the models $\mathfrak{A}_\mu$. Suppose that we have constructed the conditionally optimal models $\mathfrak{A}_{p,q}(B_i)$ for the positions B_i reached by the moves $\Psi_i = (A, B_i) \in M(A)$, with the probabilities $P(B_i) = p_i$, $Q(B_i) = q_i$, and with the smallest set $V(p_i, q_i, B_i)$ of positions that satisfy these equations. We consider a model with the tree

$$\mathfrak{A}_\mu(A) = A \cup M(A) \cup \bigcup_{\Psi_i = (A,B) \in M(A)} \mathfrak{A}_{p_i,q_i}(B_i).$$

For this model

$$P(A) = 1 - \sum_{\Omega \in M(A)} \pi(A,\Omega) \prod_{\Psi_i = (A,B_i) \in \Omega} q_i \prod_{\Psi_i = (A,B_i) \in M(A) \setminus \Omega} p_i,$$

$$Q(A) = 1 - \prod_{\Psi_i = (A,B_i) \in M(A)} p_i,$$

$$V(\mathfrak{A}_\mu) = \sum_{\Psi_i = (A,B_i) \in M(A)} V(p_i, q_i, B_i) + 1.$$

Thus, to construct the conditionally optimal models $\mathfrak{A}_{p,q}(A)$ with minimum number of positions, under the proviso that $p(A) = p$, $q(A) = q$, it suffices to consider only the conditionally optimal models $\mathfrak{A}_{p_i, q_i}(B_i)$, $\Psi_i = (A, B_i) \in M(A)$.

Of course, there may be very many such models. However, if we are content with approximate optimality the number of competing models can be decreased. We may for instance consider only those models $\mathfrak{A}_{p_i, q_i}(B_i$ for which the number of positions $V(p_i, q_i, \mathrm{B}_i)$ is minimal for $p_i \geq p$ or $q_i \leq q$. Or we may consider such models, not for all values of p and q but only for values in a more or less particularized net. Finally, disregarding the fact that the set of values of p and q for which conditionally extremal models $\mathfrak{A}_{p,q}(B_i)$ exist is discrete, we may construct by interpolation continuous functions $V(p, q, B_i)$ and apply known techniques for finding conditional extrema of functions of several variables.

Much computation is required for the determination of the values of $P(A)$ by the use of the above formula. To save computation we may limit the number of sets Ω of winning moves, by considering only those for which the corresponding probabilities $\pi(A, \Omega)$ are not too small. We may also impose a natural requirement that the values of these probabilities should be determined by a relatively small number of parameters. With these postulates the computation of the $P(A)$ is greatly simplified, and it becomes possible to make effective use of dynamic programming.

Research on methods for shortening the search by the use of various probabilistic assumptions is only just beginning, and we may expect interesting results from it. In particular, great interest attaches to a combination of the methods of non-minimax search, construction of models, and use of *a posteriori* information about results obtained earlier in the search. Also of great interest are statistical studies needed for the solution of problems about the applicability of various postulates concerning real games, about the usable evaluation functions and other parameters that define the character of a position in the tree of a model game, and also on the concrete numerical values of various probability characteristics.

Remarks on the Propriety of the Probabilistic Approach

The use of probabilistic postulates as a basis for choosing a model game and a search method may meet with principial objections, even though these postulates are supported by substantial statistical evidence in connection with such games as, for example, chess. In fact, every chess position has a true score established by a deterministic process. Also determined are the values of an arbitrary evaluation function applied in a model game. Therefore 0 and 1 are the only possible values for the probabilities of the occurrence of an elementary event, e.g. that a given position A has the true score $\text{sc}(A) = x$, or that the evaluation function $f(A) = y$, or that the value of some deterministically defined parameter $t(A) = t$. This means that the probabilities of compound events can have only the values 0 and 1, and are interrelated deterministically.

Analogous objections may be raised against many applications of probability theory, to which everyone has long since become accustomed. Clearly, the objections are not less convincing merely because of this historical acceptance. They were raised by the eighteenth century founders of probability theory. At that time, as a basis for the applications of the theory, it was argued that events governed by sufficiently complex laws behave in many relationships like random variables. A. N. Kolmogorov [20] has put forward an algorithmic approach to the concept of complexity, and has to some extent supported this argument.

We shall not give Kolmogorov's exact definition of complexity. We note only that the complexity of a finite subset Ω of a set $\mathfrak{M}$ of objects each having a finite description is defined as the length of the shortest program that will generate the subset Ω. There is a valid theorem to the effect that the values of sufficiently simple functions $\varphi(\Omega)$ defined on sufficiently complex subsets of the set $\mathfrak{M}$ will be near the correspondingly determined probabilities when Ω is taken as a set of random elements.

When we are studying the results of a minimax search in a Shannon model of depth n the finite subsets Ω are the subsets of positions in the tree of the game $\mathfrak{A}$ that satisfy the following conditions: their rank is n, their true score has a given value, and they belong to given subtrees of the Shannon model $\mathfrak{A}_\nu$. However, such subsets are not complex in the sense of Kolmogorov. To define them one need only search all positions in the game $\mathfrak{A}$, compute the true scores of all positions, and define the corresponding subsets Ω. A comparatively short program will accomplish this task.

On the other hand, in the course of developmental work on Kolmogorov's algorithmic approach to the concept of complexity, another definition of complexity has been formulated (cf. [23]). In this formulation the complexity of a program is taken as the sum of its length and the logarithm of the running time; the complexity of a subset Ω is the minimal complexity of programs that generate Ω. Thus the known algorithms for determining the true scores of positions turn out to be complex for the majority of the positions that have interested humans or game-playing programs since the invention of chess (but not for positions in chess problems requiring mate in a given number of moves). At the same time, the values of the evaluation functions and model scores for different search methods require relatively simple functions.

For the concept of complexity defined above, there are valid theorems on the values of simple functions of complex subsets, which can be formulated in the same way as the theorems based on the Kolmogorov definition. Apparently they are sufficient for the proof of the theorem on the relations between model scores and true scores, under the conditions that there exist no algorithms for defining true scores of positions more effective than those now known, and that the corresponding relationships between the values of the evaluation function and the true scores are valid.

If the first assumption holds, the second can be tested by standard statistical devices, good or bad, just as in the testing of any probabilistic hypothesis. The situation respecting the first assumption is notably worse. To validate it one must almost unavoidably inspect all algorithms displaying the desired effectiveness and prove that none will determine a true score, i.e. find a position whose score as computed by the given algorithm is incorrect (or, more exactly, one must show that for almost all tasks that may in principle be posed, the situation is exactly as described).

Thus the admissibility of the probabilistic approach to a determination of the objective quality of a program that uses a model game can be based only on an assumption about the non-existence of sufficiently simple algorithms that solve correctly the problem of computing true scores. Any method for defining complexity, if it is to be well supported by experiment, cannot make use of a highly complex exact algorithm, and the basic postulate for such a method may well turn out to be practically impossible to validate.

A similar situation has been met in discrete mathematics. A way out was found in the formulation and proof of theorems of a conditional character. A number of problems were proved to be universal: either there exists no algorithm for their solution having its running time dependent on a power of the dimension of the problem, i.e. the length of the description of a concrete instance, else such algorithms exist for all exactly posed discrete problems (a description is a word of finite length by which the object, about which the problem is posed, is uniquely established, so that if the running time is unlimited the problem can be unconditionally solved).

Conditional theorems on probabilistic models of games may be highly optimistic, e.g. theorems on the high quality of determinations of true scores based on the validity of some precisely formulated assertions about some set of objects having finite descriptions.

Nevertheless, without awaiting the formulation and proof of such theorems, we may use the probabilistic approach to the construction of new types of model games, search methods, algorithms for computing evaluation functions, and the selection of parameters defining the work of game-playing programs.

Appendix

A Sketch of Past Work on Algorithms for Games

Game programming, as one of the fundamental tasks for artificial intelligence, has drawn the attention of of many research workers, with diverse aims in regard to both the creation of game-playing algorithms and developmental work on them. The approaches employed, and the successes achieved, are correspondingly diverse.

At the dawn of the computer age, when the power of computers was not as evident as it is now, various games played by computers served more than once to demonstrate the wide scope of their powers. Examples can be found in the work of J. v. Neumann and O. Morgenstern [129] and N. Wiener [165].

C. E. Shannon devoted two papers [39, 40] to the foundations of a chess algorithm; in these he sketched an approach so general that all current chess-playing algorithms appear as implementations of it—to be sure, with modifications.

At about the same time A. M. Turing [162], devising the same scheme in somewhat more detail, undertook an experiment using a manual simulation of a chess program. This simulation resembled a game played by a rather weak player. (We omit further consideration of the results of this and other games played by chess programs. Well-annotated texts of many such games may be found in a book by M. Newborn [131]).

In 1957 a program was developed at Los Alamos that played on a reduced board having 6×6 squares: it played three games. And, finally, in 1958 A. Bernstein and others [68, 69] wrote a program that played normal chess. Although it played very poorly, it settled in the affirmative the question of whether machines could play chess.

In the sixties, many algorithms were written for various games, e.g. dominoes [30], noughts-and-crosses [31], bridge [59], and others. The general opinion was that the establishment of programs that would outplay the strongest players was only a matter of technique and a small amount of time. In a few years it became clear that the development of actually strong programs would require a substantial amount of work.

Precisely at this time there arose a legend, in our opinion not valid, but still widely accepted even now, that the development of strong algorithms for games required the participation of strong players. Among the adherents of this view are some who take a direct part in the programming—M. M. Botvinnik [7–9], H. Berliner [67], A. Steiner [32], M. Newborn—and others who take no part in programming but make up for it by their acknowledged authority in the game itself—M. Tal, R. Fischer, and others.

Disregarding the fact that in our group there are well-qualified chess players, we believe (and hope that this book is sufficiently eloquent testimony to our belief) that game programming is a matter of cybernetics and that the development of game-playing programs demands no less specific work on the algorithms than it requires knowledge of the specific game. The latter may even turn out to be detrimental, since it is not always well formalized.

A second legend says that the computer should think like and 'in the image of' the thought processes of a human, and that the thesis mentioned above is unconditionally correct. However, if our aim is to learn how to use computers for the solution of complex intellectual problems, we unjustifiably complicate the task if we first solve the problem itself, perhaps more difficult than the problem of developing computer methods.

All these legends, and the many debates arising in connection with game programming, stem from the fact that its methods have not been established nor has the topic been defined. Only one aim is clear—to create a program that will play a popular and complex game (primarily chess) better than the human player.

A notable group of theoreticians takes its departure point from experiments in the use of game programs to refine a model of human thought processes. Essentially, this work lies in the field of psychology—A. de Groot [88, 89], O. Tikhomirov [34], N. Charness [77], study the techniques of human chess players and attempt to reduce them to programs.

Botvinnik's work [7–9] sides ideologically with this trend. It is not the human thought processes in chess, but rather Botvinik's own thought processes (as he presents them) that form the basis for his construction of strong chess programs. This work is distinguished by the fact that it is accompanied by an actual program implementing the algorithm developed by Botvinnik's methods. The creative work is in its final stages, and one must wait until the program is put into play before appraising it.

A principal group of theoreticians in the field of game-playing programs constructs speculative concepts of the way humans think when engaged in playing games, and attempts to support these concepts by creating the corresponding algorithms. The majority of the investigators choose games as the intellectual activity of the computer and test the effectiveness of the general algorithm on game models.

The first such work was the (today classic) program 'General Problem Solver' of A. Newell, H. Simon, and J. Shaw [29] dating from 1958. Putting forward the hypothesis that thought is exhaustive, they wrote a program implementing the general exhaustive enumeration scheme: one interesting implementation of the scheme was a chess-playing program.

D. Michie [124] chose games as a model for working out the possibilities for machine representation of knowledge. This work has the insufficiency characterizing all work in this direction.

It is fashionable to choose a trivial game, so trivial that the advantages of the proposed concept remain unclear. (For the 25 years in which the problem of artificial intelligence has existed, the most popular model has up to now been the puzzle called the 'Tower of Hanoi'.) For Michie the model is the endgame with King and Rook against King. He is now completing his work with the more complicated endgame of King and Rook against King and Knight.

A large part of the work on game programming is devoted to methods of pattern recognition for various games. In essence these papers express a faith in the necessity of including such procedures in game-playing programs. There exists only one chess-playing program making use of pattern recognition. This is a program by A. Zobrist, F. Carlson, and K. Kalma [168–170]. It has taken part in the ACM Computer Chess Tournaments, but has invariably ended in next to last place.

Much work has been done on learning programs. Learning was applied, for instance, in a program by Kh. Brutyan et al. [12] to play simple endgames.

A. Samuel credits learning with the successes of his program for playing American checkers. [33,179]. This program is already a legend on, so to speak, a local scale. Learning in Samuel's program reduces to local variation in the weights of various factors in the evaluation function. The program carries out a complete search to a depth of nine plies with an exhaustive inspection of forced variations. In all probability, such a search depth is beyond the capabilities of even highly qualified players, and the evaluation function can vary within wide limits. For comparison, we note that the strongest chess programs use a search depth of six plies.

The work of H. Berliner [62] stands alone; he uses a so-called tactical analyzer—an algorithm that analyzes the tactical causes of failure of the plans developed by the program, and attempts to find a way around these failures. Berliner has written a program including this algorithm, but because of an ineffective representation of information, the program is a mediocre player.

J. Pitrat [137,138] developed the same idea in a program for finding combinations. He proposed a unique and very important simplification—the program knows what exactly it has to win; its task is 'find the combination winning the exchange'.

J. McCarthy [122] used chess as a model of commonsense.

R. Gadzhiev [13] used the principles of situational management to construct endgames in chess.

A group of papers has been devoted to search strategies. We note only the work of R. Banerji [49, 50], K.Church and R. Church [79], and R. Atkin [45, 46] (not to be confused with one of the authors of CHESS-4).

The remaining group seeks specific machine methods for the solution. Even Newell, Simon, and Shaw did not refrain from including in their programs various search methods, in particular the α, β-procedure, regardless of whether humans use such methods.

A characteristic feature of this direction is that the work is based on functioning programs. Unfortunately, however, not every functioning program forms a basis for serious research. A significant number of programs owe their existence to the enthusiasm of their creators, and in essence contribute nothing new to the body of research on game-playing algorithms. Nevertheless, the majority of playing programs, and especially the strong ones, are the fruit of serious scientific work.

The first example of such work, on which one would want to dwell in detail, is the Odnomastka program of A. Brudno and I. Landau [11]. Its creation was accompanied by theoretical developments. The correctness of all the methods employed for shortening the search was mathematically proved. In particular, the α, β-heuristic was first theoretically based in the course of this work. Also, their program possessed an ability—unique up till now—to set traps, i.e. to choose, from among equally valued moves, one that required of the opponent the only correct reply. Together with the high technical level of the programming, all this led to the creation of an absolutely invincible program. The only regrettable aspect of the work is that the chosen game is practically never played by humans, and the excellent quality of the program is not accorded the dignity due it.

An interesting method for realizing Shannon's scheme for chess, together with a number of effective heuristics, was applied in the chess program ITEF [2]. This publication contains the proof of an important theoretical result on the optimality of the α, β-procedure for an arbitrary game.

The majority of the American programs in the 60's [121] were devoted to the method of evaluation functions. In this method, not all moves from intermediate nodes in the search are made, but only those that are best according to some evaluation function. Usually they employ a function of the score that arises after a move to the position.The best program employing this principle is due to R. Greenblatt [104]. Moves are evaluated by an essentially different function, applied to the move itself and not to the position arising after the move.

Today the two strongest programs in the chess world, CHESS-4 [153] and KAISSA [15], are based on serious theoretical development of methods for

Table 8. North American Championship Tournament,
First International Championship for Chess Programs: 1974 List of Participants

Program	Authors	Computer	Program Development Locus	Country
1. KAISSA	Adelson-Velsky, Arlazarov, Donskoy	ICL-4/70	Institute of Control Problems Moscow	USSR
2. CHESS 4.0	Slate, Atkin	CDC-6600	Northwestern University, Illinois	USA
3. RIBBIT	Parry, Hansen, Crook	Honeywell 6000 ASEA	University of Waterloo	Canada
4. CHAOS	Ruben, Swarz, Winograd, Berman, Toikka	Univac 1110	Univac Corporation	USA
5. TECH II	Baisley	DEC PDP-10	MIT	USA
6. OSTRICH	Arnold, Newborn	Data General Nova 2	Columbia University	USA
7. FRANTZ	Wolf	Univac 494	Computer Center Graz	Austria
8. MASTER	Kent, Birmingham	IBM-370/195	Atlas Laboratory	England
9. BEAL	Beal	CDC-6400	Queen Mary College, London	England
10. FREEDOM	Barricelli	CDC Cyber-74	University of Oslo	Norway
11. TELL	Joss	HP-2100	Federal Technical University Zurich	Switzerland
12. A16CHS	Prinsen	GCS Alpha-16	International Data Systems	England
13. PAPA	Reiner, Almasi	CDC Cyber-73	Budapest University	Hungary

Table 9. North American Championship Tournament for Chess Programs: 1975

Program	Authors	Computer	Program Development Locus	Country
1. TREE FROG	Hansen, Kalnek, Crook	Honeywell 6080	University of Waterloo	Canada
2. CHESS 4.4	Slate, Atkin	CDC-6400	Northwestern University	USA
3. CHAOS	Ruben, Swartz, Winograd, Berman, Toikka	Amdahl 470	Amdahl	USA
4. OSTRICH	Newborn, Arnold	Data General Nova 2	McGill University	Canada
5. DUCHESS	Wright, Truscott	IBM 370/165	Duke University, North Carolina	USA
6. CHUTE 1.2	Valenti, Vraneshitch	IBM 370/165	University of Toronto	Canada
7. TYRO	Zobrist, Carlson	IBM 370/158	Univ. of South California	USA
8. SORTIE	Becker, Mammon, Anderson, Mann, Egan, Swingle	SIGMA-7	Bucknell University	USA
9. VITA	Marsland	IBM 360/67	University of Alberta	Canada
10. ETAOIN SHRDLU	Courtois	Nova-1200	University of Colorado	USA
11. BLACK KNIGHT	Sogg, Maltsen, Losov, Prouse	Univac 1110	UNIVAC, St. Paul, Minnesota	USA
12. IRON FISH	Buss, Mandstock	CDC Cyber-74	University of Minnesota	USA

Table 10. European Championship for Chess Programs: 1976

Program	Authors	Computer	Program Development Locus	Country
1. TELL	Joss	HP-2115	Federal Technical University Zurich	Switzerland
2. DEJA	Jan, Zagler	TR-440	Leibnitz University, Munich	FRG
3. SCHACH MB 5.6	Richter	TR-4, TR-440	Inst. of Informatics, Hamburg	FRG
4. ORWELL 3	Nitsche	UNIVAC 1106/2	University of Freiburg Computation Center	FRG
5. FISHER-SCHNEIDER	Fisher, Schneider	TR-4, TR-440	University of Stuttgart	FRG
6. SAMILL	Klein, Kruger	IBM 370/168	Bonn University Computation Center	FRG
7. PROSCHA	Hewitt, Appelrath, Franzen, Schulz, Schulz, Teschers, Fauriberger	IBM 370/158	Computing Center of the GKhK Dortmund	FRG
8. CHARLY	Keil	Siemens 4004/45	Gymnasium sv. Anny, Augsburg Computing Center	FRG

Table 11. Second World Championship Tournament for Chess Programs: 1977 List of Participants

Program	Authors	Computer	Program Development Locus	Country
1. CHESS 4.6	Slate, Atkin	CDC Cyber-176	Northwestern University, Illinois	USA
2. DUCHESS	Truscott, Wright, Jensen	IBM 370/165	Duke University, North Carolina	USA
3. KAISSA	Adelson-Velsky, Arlazarov, Donskoy	IBM 370/168	VNIISI, Moscow	USSR
4. BLITZ 5	Hyatt	Sigma 9	University of Southern Mississippi	USA
5. MASTER	Birmingham, Kent	IBM 370/168	Rutherford Laboratory, Harwell	England
6. TELL	Joss	DEC K 110	Federal Technical University Zurich	Switzerland
7. BELLE	Thompson, Condon	PDP-11	Bell Laboratories, New Jersey	USA
8. VITA	Marsland	Amdahl 470 V/6	Toronto University	Canada
9. OSTRICH	Newborn, Arnold	Supernova	McGill University	Canada
10. DARK HORSE	Ratsman	CDC-6600	Eriksen Telephone Company, Stockholm	Sweden
11. BCP	Beal	CDC-6400	Queen Mary College, London	England
12. ELSA	Zagler	TR-440	Technological University, Munich	FRG
13. CHAOS	Alexander, Macbride, Swarz, Toikka, Berman, Winograd	Amdahl 470 V/6	University of Michigan	USA
14. BLACK KNIGHT	Sogg, Prouse, Malzen, Leban, Adams	Univac 1110	Sperry-Univac, Minnesota	USA
15. CHUTE 1.2	Valenti, Vraneshitch	Amdahl 470 V/6	Toronto University	Canada
16. BS'66'76	Svets	IBM 370/168	Unaffiliated	Netherlands

shortening the search. Both programs painstakingly avoid prunings that might change the results of the search.

The theory of searching has also attracted other investigators. D. Knuth [115] and M. Newborn [133] have studied the theory of the α, β-procedure.

L. Harris [108, 109] proposed some new schemes, unfortunately not supported by the development of programs. J. Birmingham and O. Kent [71], authors of the strong West European program 'Master', developed and implemented in it several interesting heuristics.

Here we should like to eliminate some misunderstandings, essentially of a terminological nature. It is currently customary to refer to a search without pruning as a full search. We, on the other hand, continually emphasize the fact that KAISSA employs a full search, meaning in our case that the results of the search agree at all positions. In other words, the program does not make erroneous prunings.

Control by search has its shortcomings. The basic one was remarked by H. Berliner [67]—the approximate character of the model employed. When the search parameters have been chosen we have defined only an approximation of chess. For a small search depth this approximation is very rough. With an increase in the depth of search the quality of the approximation improves, but apparently not as rapidly as one would like.

We should note yet another paper in chess programming. E. Fredkin and R. Greenblatt developed at MIT a specialized microprocessor implementing the technical side of a chess game. Thus people having ideas as to how to construct a chess program but not wishing to spend years in writing a program to prove the correctness of their ideas, may test them in a short time without having to spend time in tedious routine work. Experience has shown that without a trial in an actual program it is impossible to test the fruitfulness of one's ideas.

Today the number of functioning game programs is satisfactorily large. Tournaments for the games of Go and Noughts-and-crosses are held regularly, and computer chess tournaments have been held under the aegis of IFIP. Tournaments have been held for two World Championships, seven North American Championships, and a European Championship. Tables 8–11 list the official data on the entrants in the World Championship, the 1975 North American Championship, and the European Championship.

Material pertaining to game programming is published regularly in the journals Artificial Intelligence (Holland), Man–Machine Studies (USA) and in the Proceedings of Biennial Conferences on Artificial Intelligence.

Summary of Notations

References

1. Adelson-Velsky G. M., Arlazarov V. L. Metody usileniya shakhmatnykh programm. (Methods of strengthening chess programs.)–Problemy kibernetiki, 1974, **29**, 167–168.
2. Adelson-Velsky G. M., Arlazarov V. L., Bitman A. R., Uskov A. V. and others, O programmirovanii igry vychislitelnoi mashiny v shakhmaty. (On programming computers to play chess.)–UMN, 1970, **25**, vol. 2, (152),221–260.
3. Adelson-Velsky G. M., Landis E. M. Odin algoritm organizatsii informatsii. (An algorithm for organizing information.)–DAN SSSR, 1962, **146**, No. 2, 263–266.
4. Arlazarov V. L., Futer A. L. and others. Obrabotka bol'shykh massivov informatsii na primere analiza ladeinovo endshpielya. (Processing large volumes of data for the analysis of endgames with Castles.)–Programmirovanie, 1977, No. 4, 45–54.
5. Bellman, R. Dinamicheskoe programmirovanie. (Dynamic Programming)–M.: IL, 1960.
6. Bongard M. M. Problema uznavaniya. (The recognition problem.)–M.: Nauka, 1967.
7. Botvinnik M. M. Algoritm igry v shakhmaty.–M.: Nauka, 1967.
8. Botvinnik M. M. Blok-skhema algoritma igry v shakhmaty. (Block diagram of an algorithm for chess.)–M.: Sov. Radio, 1972.
9. Botvinnik M. M. O kiberneticheskoe tseli igry. (On the cybernetic goals of a game.) M.: Sov. Radio, 1975.
10. Brudno A. L. Grani i otsenki dlya sokrashcheniya perebora variantov. (Bounds and estimates for shortening the search of variations.) Problemy kibernetiki, 1963, **10**, 141–150.
11. Brudno A. L., Landau I. Ya. Odnomastka (Programmirovanie igrovoi zadachi.) (Odnomastka: programming a game.) Problemy kibernetiki, 1965, **13**, 141–160.
12. Brutyan Kh. K., Zaslavskii I. D., Mkrtchyan L. V. O nekotorykh metodakh matematicheskogo sinteza positsionnykh strategii v igrakh. (On some methods for mathematical synthesis of positional strategies in games.)–Problemy kibernetiki, 1967, **19**, 141–175.
13. Gadzhiev R. E. Eksperimental'noe issledovanie protsessa prinyatiya resheniya chelovekom i ego modelirovanie s pomoshch'yu metoda situatsionnogo up-

ravleniya (na primere shakhmatnogo endshpielya.) (An experimental study of human decision-making process and a model of it using the method of situational control (with endgames in chess as an example.)–M., 1975, 17–27. (Trudy 4-i Mezhdunarodnoi konferentsii po isskustvennomu intellektu, **10**. (Proceedings of the 4th International Conference on Artificial Intelligence, **10**.)

14. Gol'fand Ya. Yu., Futer A. L. Realizatsiya debyutnoi spravochnoi dlya shakhmatnoi programmy. (Implementation of a reference facility for openings in chess programs.) Problemy kibernetiki, 1974, **29**, 201–210.
15. Donskoi M. V. O programme, igrayushchei v shakhmaty. (On a chess-playing program.)–Problemy kibernetiki, 1974, **29**, 169–200.
16. Evgrafov M. A., Zadykhailo I. V. Nekotorye soobrazheniya o programmirovanii shakhmatnoi igry. (Some observations on algorithms for chess.)–Problemy kibernetiki, 1965, **15**, 135–156.
17. Ershov A. P. O programmirovanii arifmeticheskikh operatorov. (On programming arithmetic operators.)–DAN SSSR, 1958, **118**, No. 3, 427–430.
18. Efimov E. I. Modelirovanie shakhmatnykh okonchanii na osnove avtomatizatsii dokazatel'stva teorem. (Modelling chess endings using automatic theorem proving.)–Izv. AN SSSR. Ser. tech. cybern., 1977, No. 2, 47–60.
19. Knut D. Isskustvo programmirovaniya dlya EVM. (D. E. Knuth. The Art of Computer Programming, v. 2, Seminumerical Algorithms.) M.: Mir, 1977.
20. Kolmogorov A. N. K logicheskim osnovam teorii informatsii i teorii veroyatnostei. (Toward the logical foundations of information theory and probability theory.)–Problemy peredachi informatsii, 1969, **5**, No. 3, 3–7.
21. Komissarchik E. A., Futer A. L. Ob analize ferzevogo endshpilya pri pomoshchi EVM. (Computer-assisted analysis of Queen endgames.)–Problemy kibernetiki, 1974, **29**, 211–210.
22. Kronrod V. A. Krestiki-noliki na pole 5×5. (Noughts and crosses on a 5×5 board.)–In: O nekotorykh voprosakh teoreticheskoi kibernetikii algoritmakh programmirovaniya. (On some problems in theoretical cybernetics, and on programming algorithms.) Novosibirsk, 1971, 185–210.
23. Levin L. A. Universal'nye zadachi perebora. (Universal search problems.)–Problemy peredachi informatsii, 1973, **9**, No. 3, 115–116.
24. Maizlin I. E. Ob odnom sposobe poiska informatsii i ego primenenii dlya realizatsii na EVM algoritma nakhozhdeniya kriticheskogo puti. (On a method for information search and an application to a computer implementation of an algorithm for finding the critical path.)–DAN SSSR, 1964, **159**, No. 4, 761–763.
25. Minskii M., Peipert S. Perseptrony. (M. Minsky, S. Papert. Perceptrons, An Introduction to Computational Geometry. MIT Press, Cambridge 1969.)–M.: Mir, 1971.
26. Neiman J. fon. K teorii statisticheskhykh igr. (v. Neumann J. Toward a theory of statistical games.)–V kn.: Matrichnye igry. (In the book: Matrix games.) M.: Fizmatgiz, 1961, 173–204.
27. Nil'son N. J. Iskusstvennyi intellekt. Metody poiska reshenii. (Nilsson N. J. Artificial intelligence. Methods of search for solutions.)–M.: Mir, 1973.
28. N'yuell A., Saimon G. GPS-programm, modeliruyushchaya protsess cheloveshcheskogo myshleniya.–V. kn.: Vychislitel'nye mashini i myshlenie. (Newell A., Simon H. GPS, a program that simulates the process of human thought. In the book: Computers and Thought.) M.: Mir, 1967, 283–301.
29. N'yuell A., Shou J., Saimon G. Programma dlya igry v shakhmaty i problema slozhnosti–V kn.: Vychislitel'nye mashiny i myshlenie. (Newell A., Shaw J., Simon H. A program for chess and the problem of complexity.) Ibid., 33–70.

30. Pervin Yu. A. Ob algoritmizatsii i programmirovanii igry v domino. (On algorithms and programs for the game of dominoes.)–Problemy kibernetiki, 1960, **3**, 171–180.
31. Srapyan Sh. O., Ter-Mikaelyan T. M. Ob odnom metode otsenki situatsii v krestiki i noliki. (On a method for assessing a position in the game of noughts-and-crosses.)–Problemy kibernetiki, 1963, **9**, 171–176.
32. Steiner A. M. BRIBIP-programma, osushchestvlayushchaya torgovlyu v bridzhe. (The program BRIBIP, implementing the bidding process in bridge.)–(Trudy 4-i mezhdunarodnoi obedinennoi konferentsii poiskustvennomu intellektu, **3**) (Proceedings of the 4th International Conference on Artificial Intelligence.)
33. Semyuel A. Nekotorye issledovaniya vozmozhnosti obucheniya mashin na primere igry v shashki.–V kn: Vychislitel'nye mashiny i myshlenie. (Samuel A. Some studies in machine learning using the game of checkers.–In: Computers and Thought) M.: Mir, 1967, 71–111.
34. Tikhomirov O. K., Poznanzkaya E. D. Issledovanie visual'nogo poiska kak sredstvo analizirovaniya evristik. (Studies of visual search as a means for analyzing heuristics.)–Voprosy psikhologii, 1966, **2**, 39–53.
35. Zermelo E. O primeneniya teorii mnozhestv k teorii shakhmatnoi igry.–V kn: Matrichnye igry. (On the application of set theory to the theory of chess.–In: Matrix games.) M.: Fizmatgiz, 1961, 167–172. (See also English translation of German paper at 5th Int. Cong. of Math. 1912.—Firbush News, **6**, July 1976, 37–42.)
36. Chikul V. M. Metod tochnechnykh baz dlya sokrashcheniya perebora variantov. (The method of point bases for shortening the search of variations.)–Voprosy kibernetiki i vychislitel'noi matematiki, Tashkent, 1966, **5**, 35–45.
37. Chikul V. M. Ob evristicheskom programmirovanii intellektual'nykh igry. (On heuristically programmed intellectual games.)–Voprosy kibernetiki i vychislitel'noi matematiki, Tashkent, 1968, **16**, 54–72.
38. Chikul V. M. Universal'naya evristicheskaya igrovaya programma. (A universal heuristic game-playing program.)–V kn: Konf. po teorii avtomatov i iskusstv. myshl. Tashkent, 1968, 11–13.
39. Shennon K. E. Mashina dlya igry v shakhmaty. (Shannon C. E. A chess-playing machine.)–V kn: Shennon K. E. Raboty po kibernetiki i teorii informatsii. (In: Shannon C. E. Papers on cybernetics and information theory.) M.: Fizmatgiz, 1956, 180–191. (Scientific American, v. 182, Feb. 1950, 48–51).
40. ———. Igrayushchie mashiny. (Game-playing machines.) Ibid, 216–223.
41. Adelson-Velsky G., Arlazarov V., Donskoy M. On the structure of an important class of exhaustive problems and on ways of search reduction for them.–Advances in Computer Chess, 1977, **1**, 1–5.
42. ———. Some methods of chess play programming.–Artificial Intelligence, 1975, **6**, 361–376.
43. ———. More commentary on the Cichelli heuristics. SIGART, 1974, **45**, 12.
44. Advances in Computer Chess, Ed. Clarke M.–Edinburgh Univ. Press, 1977, v. 1.
45. Atkin R. H., Whitten I. H. A multi-dimensional approach to positional chess.–Int. J. Man-Machine Studies, 1975, **7**, 727–750.
46. Atkin R. H., Hartston W., Whitten I. H. Fred CHAMP, Positional chess analyst.–Int. J. Man-Machine Studies, 1976, **8**, 517–529.
47. Atkin R. H. Positional play in chess by computer.–Advances in Computer Chess, 1977, **1**, 60–73.

48. Atkin R. H. Multidimensional structure in the game of chess.–Int. J. Man-Machine Studies, 1972, **4**, 341–362.
49. Banerji R. B., Ernst G. W. Changes in representation which preserve strategies in games.–In: Proc. of IJCA12, 1971, 651–658.
50. Banerji R. B. Game playing programs: An approach and an overview.–In: Theoretical approaches to non-numerical problem solving/Ed. Banerji R. B., Mesarovic M. D.–In: Proc. of the IV Systems Symposium at Case Western Reserve Univ. New York: Springer-Verlag, 1970.
51. Baylor G. W., Simon H. A. A chess mating combinations program.–Proc. of AFIPS, SJCC, 1966, **28**, 431–447.
52. Bell A. C. Techniques for playing the endgame.–Computer Weekly, 1965, April 10.
53. Bell A. G. Computer chess experiments.–See [84].
54. Bell A. G. Kalah on Atlas.–Machine Intelligence, 1967, **3**, 181–194.
55. Bell A. G. Algorithm 50: How to program a computer to play legal chess.–Computer Journal, 1970, **13**, No. 2, 208–219.
56. Bell A. G. Games playing with computers.–Allen and Unwin, London, 1972.
57. Bellman R. Stratification and control of large systems with applications to chess and checkers.–Information Science, 1968, **1**.
58. Bellman R. On the application of dynamic programming to the determination of optimum play in chess and checkers.–Proc. Natl. Academy of Sciences, 1965, **53**, 244–247.
59. Berlekamp F. Program for double dummy bridge problems.–Journal of ACM, 1963, **10**, 357–364.
60. Berliner H. J. A new subfield of computer chess.–SIGART, 1975, **53**, 20–1.
61. ———. Computer chess.–SIGART, 1975, **55**, 14–15.
62. ———. A representation and some mechanisms for a problem solving chess program.–Advances in Computer Chess. Ed. Clarke M. Edinburgh Univ. Press, 1977, 7–29.
63. ———. A comment on improvement of chess playing programs.–SIGART, 1974, **48**, 16.
64. ———. Outstanding performance by CHESS 4.5 against human opposition.–SIGART, 1976, **60**, 12–13.
65. ———. Man against machine 1974.–Firbush News, 1976, **6**, 63–70.
66. ———. Chess playing programs. SIGART, 1969, **17**, 19–20.
67. ———. Some necessary conditions for a master chess program.–In: Proc. 3rd IJCAI, Stanford, 1973, 77–85.
68. Bernstein A., Roberts M. de V. Computer vs Chess Player.–Scientific American, June 1958, **198**, 96–105.
69. Bernstein A., Roberts M. de V., Arbuckle T., Belsky M. A. A chess playing program for the IBM 704.–In: Proc. WJCC, 1958, 157–159.
70. Binet A., Psychologie des grands calculateurs et des joueurs d'échecs.–P.: Hachette, 1894. (See also Mnemonic Virtuosity: A Study of Chess Players. Genetic Psych. Monog., v. 74, 1966, 127–162. [Translation of 1893 paper in "Revue des Deux Mondes, v. 117"]).
71. Birmingham J. A., Kent P. Tree-searching and tree-pruning techniques.–See [44], **1**, 89–107.
72. Bond A. H. Psychology and computer chess.–See [84], 29–30.
73. Bond A. H. Descriptor index.–See [84], 95–112.
74. Boos G., Cooper D., Gillogly J., Levy D., Raymond H., Slate D., Smith R., Mittman B. Computer Chess programs: a panel discussion.–Proc. 1971 Annual ACM Conference 25, 97–102.
75. Bramer M. A. Computer chess: The knowledge approach.–Chess, 1976, **41**, 347–349.

76. Bratko I., Trancig P., Trancig S. Some new aspects of chess board reconstruction experiments.–In: 3rd European meeting on Cybernetics and Systems Research. Vienna 1976.
77. Charness N. Human chess skill.–See [78], 34–53.
78. Chess skill in man and machine. Ed. Frey P.–NY.: Springer-Verlag, 1977.
79. Church R. M., Church K. W. Plans, goals, and search strategies for the selection of a move in chess.–See [78], 131–156.
80. Cichelli R. J. Research progress report in computer chess.–SIGART, 1973, **41**, 32–36.
81. ———. Preliminary testing of the effectiveness of the Cichelli Depth – 2 and refutation heuristics.–SIGART, 1973, **42**, 49–52.
82. Clarke M. R. B. A quantitative study of King and Pawn against King.–See [44], **1**, 108–116.
83. ———. Some ideas for a chess compiler.–In: Artificial and Human Thinking, Elithorn and Jones (eds.), Elsevier, 1973, 189–198.
84. Computer Chess. Ed. Bell. L.: Atlas Labs, 1973.
85. Cooper R., Elithorn A. The organization of search procedures.–In: Artificial and Human Thinking. Elsevier, 1973, 199–213.
86. Coriat I. H. The unconscious motives of interest in chess.–Psychoanalytic Review, 1941, **28**, 30–36.
87. De Groot A. D. Chess playing programs.–Proc. Koninkl. Nederlands Akad. Wetensch. Ser. A-67, Amsterdam, 1964, 385–398.
88. ———. Thought and Choice in Chess. Hague, Mouton, 1965.
89. ———. Perception and memory versus thought: some old ideas and recent findings.–In: Problem Solving. ed. Kleinmuntz, B. NY.: John Wiley, 1966.
90. Dutka J., King K., Newborn M. A review of the first United States computer chess championship.–SIGART, June 1971, **28**, 14–23.
91. Eisenstadt M., Kareev Y. Toward a model of human game playing.–In: 3rd IJCAI, 1973, 458–463.
92. Elithorn A., Telford A. Computer analysis of intellectual skills.–Int. J. Man-Machine Studies, 1969, **1**, 189–209.
94. Euwe M. Computers and chess.–In: The Encyclopedia of Chess. St. Martin's Press, 1970.
95. Faster than Thought. Ed. Bowden B. V. L.: Pitman, 1963.
96. Findler N. V. Computer experiments on the formation and optimization of heuristic rules.–In: Artificial and Human Thinking. Elsevier, 1973, 177–188.
97. Frey P. An introduction to computer chess.–See [78], 54–81.
98. Gillogly J. Reader commentary on the Cichelli heuristics.–SIGART, 1973, **43**, 27–28.
99. ———. The Technology chess program.–Artificial Intelligence, 1972, **3**, 145–164.
100. Good I. J. Dynamic probability, computer chess, and the measurement of knowledge.–Firbush News, 1976, **6**, 43–62.
101. ———. The mystery of GO.–New Scientist, 1965, **427**, 172–174.
102. ———. A five year plan for automatic chess.–Machine Intelligence, 1966, **2**, 89–118.
103. ———. Analysis of the machine chess game J. Scott (White), ICI–1900 vs. R. D. Greenblatt, PDP–10.–Machine Intelligence, 1969, **4**, 267–269.
104. Greenblatt R. D., Eastlake D. E., Crocker S. D. The Greenblatt chess program.–In: Proc. FJCC, 1967, **31**, 801–810.
105. Griffith A. K. Empirical exploration of the performance of the alpha-beta tree search heuristic.–In: IEEE Trans. on Computers, Jan. 1976, 6–10.
106. ———. A comparison and evaluation of three machine learning procedures as applied to the game of checkers.–Artificial Intelligence, 1974, **5**,137–148.

107. Harris L. R. Heuristic search under conditions of error, and plan oriented play.–Artificial Intelligence, 1974, **5**, No. 3, 217–239.
108. ———. The bandwidth heuristic search.–In: Proc. 3rd IJCAI, Stanford, 1973, 23–29.
109. ———. The heuristic search: An alternative to the alpha-beta minimax procedure.–See [78], 157–166.
110.. Hayes J. E., Levy D. N. L. The world computer chess championship.–Edinburgh Univ. Press, 1976.
111. Hearst E. Man and Machine: Chess achievements and chess thinking.–See [78], 167–200.
112. Hunt E. B. Artificial Intelligence.–N.Y.: Academic Press, 1976.
113. Kent P. A simple working model.–See [84], 15–27.
114. Kister J. et al. Experiments in chess.–Journ. ACM, 1957, **4**, No. 2, 174–177.
115. Knuth D. E., Moore R. An analysis of alpha-beta pruning.–Artificial Intelligence, 1975, **6**, 293–326.
116. Kozdrowicki E. W., Licwinko J. S., Cooper D. W. Algorithms for a minimal chess player: A Blitz Player.–Int. J. Man–Machine Studies, 1961, **3**, 141–165.
117. Kozdrowicki E. W., Cooper D. W. COKO III: The Cooper–Kozdrowicki chess program.–Int. J. Man–Machine Studies, 1974, **6**, 627–699.
118. Kozdrowicki E. W. A practical application of machine learning: use of learning in an interpreter for a tree searching language.–In: Proc. IEEE Systems Science and Cybernetics Conf., San Francisco, 1968, 250–257.
119. Levy D. N. L. Computer chess–a case study on the CDC 6600.–Machine Intelligence, 1971, **6**, 151–164.
120. Malik R. Observations.–See [84], 89–94.
121. Marsland T. A., Rushton P. G. A study of techniques for game-playing programs.–Int. J. Computer Science, 1973, **4**, No. 2, 26–30.
122. McCarthy J. Programs with common sense.–In: Semantic Information Processing. MIT Press, 1968, 403–418.
123. Michie D. ALI: A package for generating strategies from tables.–SIGART, 1976, **9**, 12–15.
124. ———. King and Rook against King: historical background and a problem on the infinite board.–See [44], 30–59.
125. ———. On Machine Intelligence.–In: Edinburgh Univ. Press, 1974, 31–49, 135–142, 186–192.
126. Mittman B. A brief history of computer chess tournaments 1970–1975.–See [78], 1–33.
127. ———. First world computer chess championship at IFIP Congress. Stockholm, August 1974.–Comm. ACM, 1974, **17**, 604.
128. Mittman B., Newborn M. Results of the fourth annual U. S. computer chess tournament.–SIGART, Oct. 1973,**42**, 36–48.
129. v. Neumann J., Morgenstern O. Theory of games and economic behavior.–Princeton Univ. Press, 1944.
130. Newborn M. A summary of the third United States computer chess championship.–SIGART, **36**, Oct. 1972, 9–26.
131. ———. Computer Chess.–.: Academic Press, 1975.
132. ———. Peasant: An endgame program for Kings and Pawns.–See [78], 119–130.
133. ———. The efficiency of the alpha-beta search on trees with branch-dependent terminal node scores–Artificial Intelligence, 1977, **8**, 137–153.
134. Newell A., Simon H. A. Human Problem Solving.–Prentice-Hall, 1972.
135. ———. Computer simulation of human thinking.–Science, 1961, **134**, 2011–2017.

136. Nilsson N. J. A new method for searching problems and game playing trees.–In: Proc. of IFIP Congress, 1968, 1556–1562.
137. Pitrat J. Realisation of a general game-playing program.–In: Proc. of IFIP Congress, 1968, 1570–1574.
138. ———. A general game-playing program.–In: Artificial Intelligence and Heuristic Programming. Edinburgh Univ. Press, 1971.
139. Samuel A. L. Some studies in machine learning using the game of checkers II–Recent progress.–IBM Journal, 1967, **11**, 601–617.
140. ———. Programming computers to play games.–Advances in Computers, 1960, **1**, 165–192.
141. Scott J. J. A chess playing program.–Machine Intelligence, 1969, **4**, 255–266.
142. ———. Lancaster vs MACKHAC.–SIGART, 1969, **16**, 9–11.
143. Scurrah M. J., Wagner D. A. Cognitive model of problem solving in chess.–Cognitive Psychology, 1971, **2**, 454–478.
144. Shannon C. E. A chess playing machine.–Scientific American, 1950, **182**, 48–51.
145. Silver R. The group of automorphisms of the game of 3-dimensional tic-tac-toe.–American Mathematical Monthly, 1967, **74**, 247–254.
146. Simon H. A., Simon P. A. Trial and error search in solving difficult problems: Evidence from the game of chess.–Behavioral Sci., 1962, **7**, No. 4, 425–429.
147. Simon H. A., Barenfield M. Information processing analysis of perceptual processes in problem solving.–Psych. Rev. 1969, **76**, 473–483.
148. Simon H. A., Chase W. G. Skill in chess.–American Scientist, 1973, **61**, No. 4.
149. Slagle J. R. Artificial Intelligence; The heuristic programming approach.–N.Y.: McGraw-Hill, 1971.
150. Slagle J. R., Dixon J. K. Experiments with some programs which search game trees.–J. ACM, 1969, **16**, 189–207.
151. Slagle J. R., Bursky P. Experiments with a multi-purpose theorem-proving heuristic program.–J. ACM, 1968, **15**, No. 1, 85–99.
152. Slagle S. R. Heuristic search programs.–See [159], 246–273.
153. Slate D. J., Atkin L. R. CHESS 4.5–The Northwestern University chess program.–See [78], 82–118.
154. Smith R. C. The Schach chess program.–SIGART, 1969, **15**, 8–12.
155. Soule S., Marsland T. A. Canadian computer chess tournament.–SIGART, 1975, **54**, 12–13.
156. Tan S. T. A knowledge based program to play chess end games.–See [84], 81–88.
157. ———. The winning program.–Firbush News, 1975, **5**, 38–45
158. ———. Describing Pawn structures.–See [44], **1**, 74–88.
159. Theoretical Approaches to Non-Numerical Problem-Solving/Ed. Banerji R. B., Mesarovic M. D.–In: Proc. of the IV Systems Symposium at Case Western Reserve University. N.Y.: Springer-Verlag, 1979.
160. Thorp E., Walden W. A partial analysis of GO.–Computer Journal, 1964–65, **7**, 203–207.
161. Thorp E. A computer-assisted study of GO on M×N boards.–See [159], 303–343.
162. Turing A. M. Digital computers applied to games.–In: Faster Than Thought, 1953, 286–310.
163. Waterman D. Generalisation learning techniques for automating the learning of heuristics.–Artificial Intelligence, 1970, **1**, 121–170.
164. Weizenbaum J. How to make a computer appear intelligent: Five-In-A-Row Offers No Guarantee.–Datamation, 1962, 24–26.

165. Wiener N. Cybernetics–N.Y.: Wiley, 1948.
166. Wolf G. Implementation of a dynamic tree-searching algorithm in a chess program.–In: Proc. ACM 73, Atlanta, 206–208.
167. Zielinski G. Heuristics for computer chess.–Applied Math.and Physics Dept., Warsaw Tech. Univ. 1975.
168. Zobrist A. L., Carlson F. R. An advice-taking chess computer.–Scientific American, June 1973, 92–105.
169. Zobrist A. L. A model of visual organization for the game of Go.–SJCC, 1969, 103–112.
170. Zobrist A. L., Carlson F. R. The USC chess program.–In: Proc. ACM 1973, Atlanta, 209–212.

Subject Index